Temporal Lobe Epilepsy, Mania, and Schizophrenia and
the Limbic System

Advances in Biological Psychiatry

Vol. 8

Series Editors
J. Mendlewicz, Brussels and *H. M. van Praag,* Utrecht

S. Karger · Basel · München · Paris · London · New York · Sydney

Symposium presented during the 3rd World Congress of Biological Psychiatry, Stockholm, June 28–July 3, 1981

Temporal Lobe Epilepsy, Mania, and Schizophrenia and the Limbic System

Volume Editors
W. P. Koella, Basel and *M. R. Trimble,* London

33 figures and 21 tables, 1982

S. Karger · Basel · München · Paris · London · New York · Sydney

Advances in Biological Psychiatry

Vol. 7: Depressive Illness. Biological and Psychopharmacological Issues
J. Mendlewicz, Brussels, A. Coppen, Epsom and H.M. van Praag, Utrecht (eds.)
VIII + 244 p., 45 fig., 63 tab., 1981. ISBN 3-8055-2482-X

National Library of Medicine, Cataloging in Publication
    Temporal lobe epilepsy, mania, and schizophrenia and the limbic system
    Symposium presented during the 3rd World Congress of Biological Psychiatry, Stockholm, June 28-July 3, 1981
    Volume editors, W.P. Koella and M.R. Trimble. – Basel, New York, Karger, 1982.
    (Advances in biological psychiatry; vol. 8)
    Papers presented at the Symposium «Temporal Lobe Epilepsy, Mania, and Schizophrenia and the Limbic System», sponsored by Ciba-Geigy läkemedel, Sweden.
    1. Bipolar Disorder – congresses 2. Epilepsy, Temporal Lobe – congresses 3. Limbic System – pathology – congresses 4. Schizophrenia – congresses I. Koella, Werner P. (Werner Paul), 1917 II. Trimble, Michael R. III. Ciba-Geigy läkemedel IV. Symposium «Temporal Lobe Epilepsy, Mania, and Schizophrenia and the Limbic System (1981: Stockholm, Sweden) V. World Congress of Biological Psychiatry (3rd: 1981: Stockholm, Sweden) VI. Series
    W1 AD44 v. 8/WL 385 T288 1981
    ISBN 3-8055-3494-9

Drug Dosage
    The authors and the publisher have exerted every effort to ensure that drug selection and dosage set forth in this text are in accord with current recommendations and practice at the time of publication. However, in view of ongoing research, changes in government regulations, and the constant flow of information relating to drug therapy and drug reactions, the reader is urged to check the package insert for each drug for any change in indications and dosage and for added warnings and precautions. This is particularly important when the recommended agent is a new and/or infrequently employed drug.

All rights reserved
    No part of this publication may be translated into other languages, reproduced or utilized in any form or by any means, electronic or mechanical, including photocopying, recording, microcopying, or by any information storage and retrieval system, without permission in writing from the publisher.

©   Copyright 1982 by S. Karger AG, P.O. Box, CH-4009 Basel (Switzerland)
    Printed in Switzerland by Werner Druck AG, Basel
    ISBN 3-8055-3494-9

Contents

Acknowledgements VI
Trimble, M.R.; Koella, W.P. (London/Basel): Introduction – Reasons and Aims of this Symposium VII
Trimble, M.R. (London): Phenomenology of Epileptic Psychosis: A Historical Introduction to Changing Concepts 1
Koella, W.P. (Basel): The Functions of the Limbic System – Evidence from Animal Experimentation 12
Andersen, P. (Oslo): Synaptic Control and Modulation of Hippocampal Neurones 40
Stevens, J.R. (Washington, D.C.): Risk Factors for Psychopathology in Individuals with Epilepsy 56
Sherwin, I. (Boston, Mass.): The Effect of the Location of an Epileptogenic Lesion on the Occurrence of Psychosis in Epilepsy 81
Trimble, M.R.; Perez, M.M. (London): The Phenomenology of the Chronic Psychoses of Epilepsy 98
Heath, R.G. (New Orleans, La.): Psychosis and Epilepsy: Similarities and Differences in the Anatomic-Physiologic Substrate 106
Post, R.M.; Uhde, T.W.; Ballenger, J.C.; Bunney, W.E. (Bethesda, Md.): Carbamazepine, Temporal Lobe Epilepsy, and Manic-Depressive Illness 117
Koella, W.P.; Trimble, M.R. (Basel/London): Epilogue 157

Subject Index 163

Acknowledgements

This volume contains the papers presented at the Symposium 'Temporal Lobe Epilepsy, Mania, and Schizophrenia and the Limbic System' during the 3rd World Congress of Biological Psychiatry, June 28–July 3, 1981. The Symposium and the printing of this volume were kindly sponsored by Ciba-Geigy, Läkemedel, Sweden. We are indebted to Mr. *Eric Wosse,* Product Manager, Ciba-Geigy, Läkemedel for his fine support of our endeavor. Miss *Jacqueline Baud's* never tiring help with editorial and secretarial chores is warmly acknowledged.

October 1981
Werner P. Koella
Michael R. Trimble

Introduction – Reasons and Aims of this Symposium

Michael R. Trimble[1], Werner P. Koella[2]

[1] The National Hospitals for Nervous Diseases, London, England; [2] Friedrich Miescher Institute, Basel, Switzerland

Schizophrenia, mania (the 'high' phase of manic-depressive illness) and temporal lobe epilepsy (the complex partial seizures) usually are looked upon as three separate diseases. Still there can be little doubt that any two, if not all three, of these nosological entities exhibit a good deal of similarity in their overt psychopathological manifestations, or – to put it differently – reveal an often considerable overlap in their symptomatological profile. This is true mainly for the 'acute' phases of these afflictions, and sometimes it is extremely difficult, if not impossible, to come to a diagnosis initially – which even then, later after lengthy and involved examinations, may still turn out to be wrong.

Furthermore, patients with epilepsy, in particular those with temporal lobe epilepsy, in the interictal interval often reveal psychiatric symptoms which may resemble mania, or assume the presentation of a florid, schizophrenia-like psychosis. Associations between such psychopathologies and epilepsy have been observed for several hundred years, but have only in this century, with the introduction of newer nosological systems in psychiatry and epilepsy, become crystallized. The early writers discussed concepts of the 'epileptic personality', an idea that mutated with time, and which recently has become once again the centre of much discussion. Originally characterized by a variety of adjectives, mostly pejorative, such a personality was seen as the inevitable consequence of having fits. It then assumed, in the 'psychosomatic era' of the midcentury, the role of a predisposition, the epileptic personality, on account of various drives, motivations and conflicts, being prone to epileptic seizures. This was followed by a period of rejection of the whole concept, a position still argued by some today, although renewed interest in the idea of an organic personality change which occurs secondary to chronic temporal lobe lesions, of which complex partial seizures represent the paradigm, has been seen when more sophisticated methods of investigating such changes were developed. Thus the idea that patients with epilepsy suffer from not only seizures, but also from aberrant behaviour traits, on account of their brain lesions, is still actively being investigated, and attempts are being made to define which patients with which lesions are most prone to which disorders. The concept of personality change due to temporal lobe lesions has, however, broadened to include other disorders, especially the schizophreniform presentations and its relationship to similar disorders in non-epileptic populations.

In addition to these clinical relationships, in the past 30 years it has been possible to define overlap between at least pairs of the above-mentioned triad in therapy in the sense that identical drugs are used to combat any two of the three afflictions. Neuroleptics have been and are used in the treatment of psychosis and mania, and – as found quite recently – carbamazepine, the drug of choice for temporal lobe epilepsy, may turn out to be useful in the treatment of mania.

If we accept the notion that a particular symptom – or most generally speaking, a piece of aberrant 'internal' and/or 'external' behaviour – is the manifestation of a specific aberrant neural performance in the system(s) that form(s) the organizational basis of that behaviour, we must conclude that the three disease entities overlap also in their pathogenesis or neuropathology. Several of the participants of this symposium and many others, on a variety of occasions, have pointed out that this (at least partial) common pathology involves structures within the limbic system including the hippocampus, the amygdala, the septum and basal limbic forebrain such as the nucleus accumbens and the olfactory tubercle and some of the interconnecting pathways.

On account of these relationships we thought it timely to bring together clinicians and basic researchers to discuss these issues. As a background for a better understanding of common pathophysiological features for temporal lobe epilepsy, psychosis and mania, Drs *Koella and Andersen* will review some of the more pertinent aspects of the gross psychophysiology of the limbic system, and the 'microscopic' physiology of the hippocampus, respectively. Their data offer important information for the physiological background of drug therapy.

Some clinicians from the field of neuropsychiatry will discuss and define again the questions of coexistence and overlap of symptoms in temporal lobe epilepsy and psychosis, and the 'therapeutic' overlap between temporal lobe epilepsy and mania. Dr. *Trimble,* Dr. *Sherwin* and Dr. *Stevens* will – from their particular point of view – present data from their own research and from the literature, that hopefully will allow further crystallization of these complex issues.

These chapters explore those groups of patients most susceptible to develop behaviour disturbances in epilepsy, highlighting some of the methodological difficulties of past studies. The issue of what type of epilepsy is related to what type of psychosis is elaborated on, and some recent data emphasizing the association between left-sided or dominant temporal lobe lesions and schizophreniform psychosis is presented.

The link between affective, in particular manic-depressive disorders and epilepsy is discussed in detail by *Post.* The possible role of kindling as a model for the development of psychosis in patients is introduced, as well as a theoretical basis for the use of certain anticonvulsants in the treatment of non-epileptic disorders in which limbic system dysfunction may occur is discussed.

When the two editors started to organize the workshop presented at the Third World Congress of Biological Psychiatry in Stockholm in July 1981, on which this book is based, Dr. *Robert Heath* of Tulane University (New Orleans, La.) was also invited to present his views on the role of the limbic system in the pathogenesis of psychosis. Unfortunately, Dr. *Heath* was unable to attend the meeting, but in view of his great contributions to this field the editors wished him to participate in this venture and thus have included a chapter by him.

The results presented support the idea of a link uniting psychosis, mania and epilepsy. We hope that the material assembled in this book will be of value, not only in introducing new ideas and avenues for further research, but also in emphasizing the biological basis for human behaviour and the role of somatic structures, especially the limbic part of the brain in the pathogenesis of behaviour disorders. 'The myth of mental illness' will some day disappear and we would have liked to have played a small part in that process.

Phenomenology of Epileptic Psychosis: A Historical Introduction to Changing Concepts

Michael R. Trimble

Consultant Physician in Psychological Medicine, The National Hospitals for Nervous Diseases, London, England

Before exploring the phenomenology of the psychoses of epilepsy, it is important to consider the whole question of classification in psychiatry. In order to do this the texts of our forefathers must be examined and in doing so it must be appreciated not only that discussions and debates regarding phenomenology have gone on for many years, but in addition that the state reached today is part of an ongoing process of evolution in our ideas regarding psychiatric nosology.

The most important and substantial advances in this field started in the 18th century. It was *William Cullen,* Professor of Medicine and Practise of Physic at Edinburgh, who laid down a classification of all diseases based on pathological theory and clinical knowledge of the time. He incorporated into his system the earlier classifications of such writers as *Sauvages* and *Linné,* and felt that his scheme was an improvement on those of *Boerhaave, Stahl* and *Hoffman* before him. Four classes of disease were recognised, which were pyrexiae, locales, cachexia (wasting diseases), and the neuroses. The latter was subdivided again into four orders which were comata (including apoplexy and paralysis), adynamia (including syncope and hypochondriasis), spasmi (including epilepsy and hysteria), and vesaniae or the insanities proper. The latter interestingly enough were reserved for 'lesions of our judging faculty' and included delirium which was of two kinds – that combined with pyrexia, and that without pyrexia, which he referred to as 'insanity'. This order was again subdivided into amentia, melancholia, mania, and oneirodynia, which was disturbed imagination during sleep [4].

The terms 'mania' and 'melancholia' continued to be used in many classifications, and authors such as *Beddoes,* physician from Clifton near Bristol, drew together these disorders recognising their often concurrence. The French physician *Esquirol* made the clear distinction between amentia, in which intellectual faculties never developed, and dementia or loss of those faculties due to disease. He also introduced in his terminology, monomania, a term reserved for partial insanity [5].

Heinroth, Professor of Psychological Medicine at the University of Leipzig, divided mental diseases into three classes: those related to the feelings, those related to the understanding, and those related to the will. He resurrected the term 'paranoia', which had earlier been used by *Vogel* as being equivalent to *William Cullen's* vesania. *Heinroth's* classification was essentially psychological, and was taken up in England by *Pritchard* who defined moral insanity as one class of disorder, and intellectual insanity as another, comprising monomania, mania, and dementia. By monomania he was also referring to the concept of partial insanity.

These terminological distinctions underwent marked revision with the successive attempts at classification by Professor *Kraepelin,* from the University of Munich, who, cast from the same mould as *Sydenham,* recognised a certain number of disease entities in his patients, each of which had a distinct cause, form, outcome and pathology. His classification included melancholia and manic-depressive conditions, which he distinguished from dementia praecox, 'a series of states, the common characteristic of which is a peculiar destruction of the internal connections of the psychic personality'. The symptoms of this disorder were variegated and amongst the psychic ones included hallucinations and delusions, which were associated with 'emotional dullness'. Paranoia, while undergoing various positions in his classification, denoted an insidious development of unshakable delusions in a relatively intact personality, set apart from dementia praecox [13].

The next major development was the introduction by *Bleuler* of the term 'schizophrenia', a disease which runs a chronic course 'and is characterised by a specific kind of alteration of thinking and feeling, and of the relations with the outer world that occur nowhere else'. He defined so-called basic symptoms, which included disturbances of association and affect, and ambivalence. However, sensation, memory and orientation, he felt, were normally intact. For him, hallucinations, delusions, and the gamut of other symptomatology noted were defined as accessory symptoms, and he accepted a sub-classification into paranoid, catatonic, hebephrenic and simplex varieties. However, under his classification of

Historical Introduction

schizophrenia, he included 'many atypical melancholias and manias of other schools (especially nearly all hysterical "melancholias and manias"), most hallucinatory confusions, much that is elsewhere called amentia, a part of the forms consigned to delirium acutum, motility psychosis of Wernicke, primary and secondary dementias without special names, most of the paranoias of the other schools, especially all the hysterically crazy, nearly all incurable "hypochondriacs", some "nervous people", and compulsive and impulsive patients' [1].

History at this point is particularly important, since it may be no coincidence that *Bleuler* succeeded *Auguste Forel* as Director of the Burghölzli Hospital, and one of several other physicians that also trained there was *Adolf Meyer*, who later went on to have a profound influence in American psychiatry. *Meyer's* own viewpoint will not be discussed here, save to point out that his psychobiology led to the introduction of 'reaction types' in diagnostic formulations, and the consideration of the individuality of each particular patient and his symptomatology in diagnostic formulation, excluding from consideration therefore the disease entities defined by *Kraepelin*.

Meanwhile in Europe, *Schneider* defined symptoms of first and second rank in schizophrenia. In trying to characterise constantly recurring features of various psychopathological states, he defined first-rank symptoms as those which 'always signify schizophrenia', although he recognised that the diagnosis could be made in the absence of these symptoms. These included the hearing of one's thoughts spoken aloud inside the head, the hearing of voices commenting on one's actions, experiences of body influence, thought withdrawal and other forms of thought interference, thought diffusion, delusional perception, and in addition, everything in the spheres of feeling, drive and volition which the patient experiences as imposed upon him or influenced by others [19].

These criteria laid down by *Schneider* have become adopted by the majority of British psychiatrists for diagnostic purposes when assessing schizophrenia, while in the United States, *Feighner* et al. [7] laid down criteria which included the following: a chronic illness with at least 6 months of symptoms without return to a premorbid level of psychosocial adjustment, and absence of a period of depressive or manic symptoms sufficient to qualify for a diagnosis of affective disorder. The patient must have either delusions or hallucinations without significant perplexity or disorientation, or verbal production which makes communication difficult on account of lack of logical or understandable organisation of thought. In addition three of the following must be present: (1) single; (2) poor premorbid social adjustment or work history; (3) family history of schizophrenia; (4) absence of alcoholism or drug abuse within 1 year of onset of the psychosis, and (5) onset of the illness prior to the age of 40.

Although authors such as *Slater* in the 1930s and *Lewis* in the 1940s commented on diagnostic differences between the United States and Great Britain, in the 1960s an Anglo-American research team known as the US-UK Diagnostic Project was set up to study differences in diagnostic practice between the two countries, the results of which have been widely reported by *Wing* et al. [25]. In the initial studies, in which samples of patients drawn from State hospitals in New York and Greater London were compared, significant differences were noted between the hospital diagnoses for almost all major categories of psychiatric illness. However, when attempts were made to tighten up diagnostic criteria, in particular using the International Classification of Diseases, concordance between the two hospitals improved. With regard to the hospital diagnoses, over 60% of the New York patients received a classification of schizophrenia, compared with less than 30% of the London sample. In contrast, depressive psychoses, mania, neuroses and personality disorder were used more frequently in London, making up nearly 50% of the total.

Following these early studies, psychiatrists were then asked to rate the same videotape material, and their diagnoses were compared. Marked disagreements were noted between the Americans and the British, again in particular with reference to schizophrenia. These studies indicated how the American concept of schizophrenia covered substantial parts of what in Britain would be regarded as depressive, neurotic, or personality disorders, and most of what British psychiatrists diagnosed as mania.

These studies demonstrated the necessity for improving reliability for psychiatric diagnoses, particularly between different countries. Nowhere can this be more important than in the field of psychiatric research, in particular in neuropsychiatry when attempts are being made to correlate changes in brain disturbance with consequent behaviour patterns. This need has led to the development of structured interview techniques for gathering psychopathological data which will be described later (pp. 90–105) in more detail. However, it is first important to consider the phenomenology of the psychoses associated with epilepsy.

Epileptic Psychosis

Traditionally it is possible to divide the epileptic psychoses into ictal and inter-ictal categories, the former being directly associated with the abnormal electrical activity of the ictus and leading to a variety of behaviour patterns which are associated with characteristic disturbances of the electroencephalogram. These are only briefly mentioned here, since no adequate studies of mental states in these conditions exist, most reports being at the anecdotal level. It should, in passing, however be noted that cases have been described of paranoid delusional states with visual and auditory hallucinations associated with generalised absence seizures of late onset, and several authors have discussed the phenomenon of complex-partial seizure status when attacks may last from hours to days during which quasi-psychotic behaviour has been detected [16]. More commonly occurring however are post-ictal psychotic states which are ushered in by major-generalised or complex-partial seizures in which auditory hallucinations, paranoid delusions, and disturbances of affect, including suicidal attempts, have been described.

With regard to the phenomenology of the chronic inter-ictal psychoses, in patients who do not show clouding of consciousness, there are descriptions of such states dating back many years. *Willis,* in the 17th century, noted that epileptic patients, during their paroxysm and afterwards 'suffer a severe loss of memory, intellect and phantasy' [24]. *Griesinger* [10] too discussed chronic mental changes, noting 'that a very great number of epileptic patients are in a state of chronic mental disease even during the intervals between their attacks'. He quoted the work of *Esquirol,* who noted that, at the Salpêtrière, of 385 epileptic women, 46 were hysterical, 30 were maniacs, 12 were monomaniacs, 8 were idiots, 145 were dements, and 50 were weak in memory or had exalted ideas. He observed that melancholia, especially with a suicidal tendency, occurred in many patients with epilepsy. *Bouchet and Cazauvieilh* [2] published a pathological report, noting an association between temporal lobe pathology and insanity, the main involvement being in the cornes d'ammon. *Falret* [6] classified epileptic psychiatric symptomatology into three types: the peri-ictal, the inter-ictal, and the long-term insanities. The latter were described as 'those phenomena of longer duration constituting true madness, whose onset should be described as either associated with or independent of any seizural manifestations'.

There was thus in the last century a growing literature on the links between epilepsy and chronic psychoses. In spite of these observations, the majority of authors to discuss insanity in epilepsy in the last century, and well into this century, concentrated on peri-ictal rather than inter-ictal states. Indeed, a number of authors advocated principles of biological antagonism, in which psychosis and epilepsy were rarely seen together, leading to the later concepts of forced normalisation by *Landolt* and its clinical equivalent, namely an inverse relationship between symptoms of psychosis and the frequency of epileptic seizures.

One of the early exhaustive accounts of the relationship between schizophrenia and epilepsy was given by *Glaus* [9] from the University Clinic in Zürich. He, and *Gruhle,* noted that the symptoms seen in schizophrenia associated with epilepsy were the same as those seen in schizophrenia without epilepsy. Referring to *Bleuler's* observations that in cases of epilepsy where delusions of persecution and auditory hallucinations are constantly present one must assume a diagnosis of schizophrenia, he felt that schizophrenia-like symptomatology occurring in epileptic patients must lead to the diagnosis of a combination of the two disorders, especially if the psychiatric abnormalities were shown to be in a setting of unaltered consciousness. He felt that hereditary predispositon could account for such combinations, and that the schizophrenia tended to develop at a time when the epileptic seizures decreased or disappeared altogether.

In the United Kingdom there have been reports since the turn of the century on epileptic psychosis developing in asylums. A number indicate that the commonest mental disorder in epilepsy was melancholia, although *White* [23], describing patients from the City of London Asylum in 1900, noted: 'moral decadence yet often religious fervour ... frequently they have delusions of a religious nature and of persecution, with hallucinations of one or more of the special senses of the familiar types. These delusions and want of self-control often cause homicidal acts.' *Savage* [18] also noted: 'a certain number of insane epileptics develop definite delusions and ideas of persecution. Hallucinations are common, occurring as aura in some cases.'

These chronic inter-ictal psychoses of epilepsy resumed renewed importance in the English literature in the 1950s with reports of such authors as *Hill* [11] and *Pond* [17]. The former described a paranoid hallucinatory psychotic state in association with temporal lobe epilepsy. *Pond,* agreeing with *Marchand and de Ajuriaguerra* [15], felt that manic-depressive psychosis was quite rare in epilepsy, its alleged association arising from too loose a diagnosis being applied to minor disturbances of affect with alteration of consciousness. However, with regard to the relationship between schizophrenia and epilepsy, again he commented on the chronic paranoid hallucinatory states noted in chronic epilepsy, which he defined as 'a definite clinical entity'. In his cases from the Maudsley Hospital, all patients had temporal lobe epilepsy with complex auras, and the epileptic attacks were noted to begin some years before the psychotic symptoms emerged. He also noted an inverse relationship between the diminution of seizures and the psychotic tendency, and described the symptoms as 'paranoid ideas which may become systematised, ideas of influence, auditory hallucinations often of a menacing quality, and occasionally frank thought disorders with neologisms, condensed words and inconsequential sentences'. Religious colouring of these paranoid ideas was common, and he commented on the warm affect of such patients which stood out in contrast to the affective changes of process schizophrenia.

The most comprehensive study was carried out in 1963 by *Slater and Beard* [20]. They examined 69 fresh cases of epileptic psychosis, and argued on statistical grounds that the two did not occur together by chance. The symptomatology included delusions in clear consciousness, which were noted in all but 2 patients. Primary delusional experiences, supposedly characteristic of schizophrenia, were seldom observed, although mystical delusions with religious overtones were very common. Passivity feelings were prominent findings, and were often closely connected with systematised ideas of persecution. Special powers were claimed by many patients, such as the ability to heal people by looking at them, or being able to see through walls or split atoms.

The majority of patients also had hallucinatory experiences, often visual and complex. However, the commonest hallucinations were auditory, frequently persecutory. Voices talking about the patient in the third person, and those commenting on the patients' actions or repeating their own thoughts were likewise relatively common. Thought disorder was shown in half the patients, usually 'a relative incapacity to handle abstract concepts'. Thought blocking and neologisms were also commented upon, as were sentences which were never finished, or showed disturbed syntax.

Affective disturbances of some kind were shown by all their patients, usually in the form of periodic moods of depression or irritability. These were often short-lived and severe, and 17 of their patients had attempted suicide on one or more occasions. So-called flattened affect was noted in 28 patients, but more commonly affect remained warm, in contrast to process schizophrenic patients. Manneristic behaviour was noted in many, although catatonic phenomena were rare. Other features defined in the patients of *Slater and Beard* [20] of importance to the clinical presentation were a negative family history for schizophrenia, an absence of schizoid premorbid personality in the patients' development, a high incidence of temporal lobe abnormalities on neurological examination and investigation, a high incidence of organic lesions, and in some patients a diminution of seizure frequency with the onset of the psychosis. They in addition commented on the apparent positive correlation between the age of onset of epilepsy and the age of onset of psychosis, with a mean interval of 14 years, suggesting from their data that an aetiological relationship may exist between the fits and the psychosis. On the whole, however, they favoured the view that the fits themselves did not cause the psychosis, but that the cause of the epilepsy was also the cause of the psychosis.

The next important paper came from *Flor-Henry* [8] in 1969. In a controlled but retrospective investigation of psychoses in temporal lobe epilepsy, he examined the case records of 50 psychotic patients who had been admitted to the Maudsley Hospital over a 15-year period. In his categories, basically Kraepelinian, the diagnosis of schizophrenia hinged on the presence of thought disorder and disturbances of affect with or without secondary symptoms such as hallucinations. In contrast, manic-depressive states were characterised by euphoric or depressive alterations of mood, exhibiting periodicity and leaving the personality intact between phases. As controls he used patients that also had temporal lobe epilepsy, in whom psychiatric disturbances were absent. The importance of this paper was not phenomenological however, but that he introduced the concept of lateralisation into the discussion of psychosis. Thus, 42% of his cases were diagnosed as schizophrenic, 22% as schizoaffective, 18% as manic-depressive, and 18% as confusional. Manic-depressive symptoms were related more to right-sided foci, and schizophrenic to left-sided or bilateral foci. In addition, he confirmed the emergence of psychosis was related to a reduced frequency of psychomotor seizures.

Other authors since this time have been less concerned with phenomenological issues and more with confirming certain neuropathological and neurophysiological relationships to psychosis. *Bruens* [3] examined 19 patients with epilepsy and psychosis, some from the Hans Berger Clinic, and others from the Utrecht University Psychiatric Clinic, and noted the symptomatology to be as follows: 9 patients had paranoid syndromes with delusions more or less systematised, 5 had psychosis with marked mental regression and transient paranoid symptoms, 2 had schizophrenia-like psychosis with thought disorder and affective disturbance, and 3 had relatively short-lived confusional states with hallucinations. Hallucinations occurred in 15 out of the 19 patients and were auditory in 12. They usually consisted of voices which were heard almost continuously, without clouding of consciousness. Religious and sexual themes were both common contents to these phenomena, and visual hallucinations were reported in 7 cases. Delusions were seen in 15 of the patients, in 9 these were delusions of reference, grandiose delusions being the next most frequent reported. In none of these cases was manic-depressive symptomatology a predominant syndrome constituent, although it is important to note that delusions of guilt occurred in 4. Temporal lobe epilepsy was again the commonest type of epilepsy in his patients, especially in the paranoid and regressive group. In his opinion however 'none of these psychoses fulfilled the strict criteria set for the diagnosis of schizophrenia'. There was 'no praecox feeling'.

Kristensen and Sindrup [14] examined 96 patients with paranoid hallucinatory psychosis and epilepsy and confirmed many of the above earlier findings, although noted that many psychotic patients had automatisms when compared to controls. They did not discuss the phenomenology, preferring to use the term 'epileptic psychosis' to discuss their patients. *Jensen and Larsen* [12] carried out a follow-up study of 74 patients with drug-resistant temporal lobe epilepsy who had unilateral anterior temporal lobectomy performed between 1960 and 1969. Many patients demonstrated behavioural disturbances, but 9 became psychotic during the follow-up period, not having been psychotic before operation. Again the phenomenology is not clearly defined, schizophrenia-like psychoses being characterised as 'psychosis with paranoid delusions'.

In addition to these studies, further evaluations have been carried out by such writers as *Taylor* [21] and *Toone and Driver* [22]; the contributions of Dr. *Sherwin* are given elsewhere in this volume. With the exception of the work of *Toone and Driver* and our results from the National Hospital [this volume], there has been little or no systematic attempt, using modern techniques, to precisely quantify phenomenology of the epileptic psychoses, and compare their presentation to similar disorders occurring in non-epileptic patients. However, from the literature, certain facts emerge. First, a chronic psychotic disorder has been noted in some patients with epilepsy for a very long time. Secondly, with the terminology we now use this often has a 'schizophreniform' presentation. Thirdly, some relationship to temporal lobe seizures has been suggested by several authors. Further work in this field is clearly required, and it is hoped that evidence presented here and in later chapters will stimulate others to investigate these problems.

References

1 Bleuler, E.: Textbook of psychiatry (MacMillan, New York 1923).
2 Bouchet, M.; Cazauvieilh, M.: De l'épilepsie considérée dans ses rapports avec l'aliénation mentale. Archs. gén. Med. *9:* 510–542 (1826).
3 Bruens, J.H.: Psychosis in epilepsy. Psychiatria Neur. Neurochir. *74:* 175–192 (1971).
4 Cullen, W.: Nosology; or a systematic arrangement of diseases, by classes, orders, genera, and species; with the distinguishing characters of each, and outlines of the systems of Sauvages, Linnaeus, Vogel, Sagar and Macbride (Creech, Edinburgh 1800).
5 Esquirol, E.: Des maladies mentales (Paris 1838).
6 Falret, J.: De l'état mental des épileptiques. Archs. gén. Méd. *16:* 661–679 (1860).
7 Feighner, J.P.; Robins, E.; Guze, S.B., et al.: Diagnostic criteria for use in psychiatric research. Archs gen. Psychiat. *25:* 57–63 (1972).
8 Flor-Henry, P.: Psychosis and temporal lobe epilepsy. Epilepsia *10:* 363–395 (1969).
9 Glaus, A.: Über Kombinationen von Schizophrenie und Epilepsie. Z. ges. Neurol. Psychiat. *135:* 450–500 (1931).
10 Griesinger, W.: Mental pathology and therapeutics (New Sydenham Society, London 1857).
11 Hill, D.: Psychiatric disorders of epilepsy. Med. Press *229:* 473–475 (1953).
12 Jensen, I.; Larsen, J.K.: Mental aspects of temporal lobe epilepsy. J. Neurol. Neurosurg. Psychiat. *42:* 256–265 (1979).
13 Kraepelin, E.: Dementia preaecox and paraphrenia (Livingstone, Edinburgh 1919).
14 Kristensen, O.; Sindrup, E.H.: Psychomotor epilepsy and psychosis. Acta neurol. scand. *57:* 361–370 (1978).
15 Marchand, L.; Ajuriaguerra, J. de: Epilepsies (Desclee De Broviser, Paris 1948).
16 Markland, O.N.; Wheeler, G.; Pollak, S.: Complex partial status epilepticus. Neurology *28:* 189–196 (1978).
17 Pond, D.A.: Psychiatric aspects of epilepsy. J. Indian med. Prof. *3:* 1441–1451 (1957).
18 Savage, G.H.: Some relationships between fits and mental disorder. Practitioner *89:* 1–10 (1912).
19 Schneider, K.: Primäre und sekundäre Symptome bei der Schizophrenie. Fortschr. Neurol. Psychiat. Grenzgeb. *25:* 487–490 (1957).
20 Slater, E.; Beard, A.W.: The schizophrenia-like psychoses of epilepsy. Br. J. Psychiat. *109:* 95–150 (1963).
21 Taylor, D.: Factors influencing the occurrence of schizophrenia-like psychoses in patients with temporal lobe epilepsy. Psychol. Med. *5:* 249–254 (1975).
22 Toone, B.K.; Driver, M.V.: Psychoses and epilepsy; in Parsonage, Aspects of epilepsy (MCS Consultants, 1980).

23 White, E.W.: Epilepsy associated with insanity. J. ment. Sci. *46:* 73–79 (1900).
24 Willis, T.: Thomas Willis's (1621–1675) Oxford Lectures by K. Dewhurst (Sandford Publications, Oxford 1980).
25 Wing, J.K.; Cooper, J.E.; Sartorius, N.: The description and classification of psychiatric symptoms (Cambridge University Press, London 1974).

M.R. Trimble, MRCP, FRCPsych., Consultant Physician in Psychological Medicine,
The National Hospitals for Nervous Diseases, Queen Square,
London WC1 3BG (England)

The Functions of the Limbic System – Evidence from Animal Experimentation

Werner P. Koella

Friedrich Miescher Institute, Basel, Switzerland

Introduction, Concepts, Anatomy, Methodological Principles

The present paper describes, in an of necessity sketchy fashion, the results of a variety of 'classical' as well as more recent psychoneurophysiological experiments whose results, in one way or the other, furthered our insight into the functional significance of the various structures of the limbic system. For a long time, indeed, this so-called 'rhinencephalon', which has little physiological relevance for the organization and integration of the olfactory functions, was 'terra incognita' for the neuro- and psychophysiologists. Although quite impressive by its mere volume, this 'borderline system' (the 'grand lobe limbique' according to *Broca*) – including such cortical and nuclear structures as the hippocampus, the cingulate cortex, the amygdala, the septal and preoptic areas, the nucleus accumbens, the olfactory tubercle, parts of the diencephalon, the mesencephalon and all connecting pathways such as the medial forebrain bundle – has been neglected in the otherwise intensive search of the neurobiologist for 'brain function'. Then in 1937 *Papez* [96] suggested, yet still without much experimental evidence, that a limbic 'circular system' consisting of the mamillary bodies, the Vicq d'Azyr tract, the anterior thalamus, the cingulum, the hippocampus, and the fornix, may serve the organization of such 'psychic' functions as mood and drive. But only the investigations of the last 40 or so years – initiated by *Klüver, Nauta, MacLean*, just to name a few – yielded those results that enable us today to offer at least some suggestions as to what this limbic lobe does for our behavior in general, and, more specifically, for our feelings and drives, for our memory and possibly for our cognitive functions; and, in addition, to make some at least educated guesses as to what may go wrong in this 'mystical' lobe if we suffer from psychosis, psychomotor epilepsy, or from melancholia.

Before we begin to describe and discuss these experimental results, it may not be superfluous, indeed it may help later interpretations and comparative considerations, if we start out by shortly characterizing the principal experimental neuropsychophysiological procedures available. Quite generally, if we intend to learn about the functional role played by a particular structure or 'system' in the organization of a particular behavior[1], we always use one of the three experimental approaches of classical neuropsychophysiology: the lesion, the stimulation, and the recording technique.

With a *lesion* we reduce or completely eliminate the activity in a particular structure, system or subsystem of the CNS and investigate, often in longitudinal studies, the qualitative or quantitative changes of a particular behavior or of a set of behaviors. This procedure offers a first insight into the necessary – by no means sufficient – part played by the (lesioned) structure for the making of the behavior under investigation. In the era of classical neurophysiology the lesion was made usually in a 'diffuse', nondifferentiating manner by (local) electrolysis, with the knife (severing pathways or the whole neuraxis), or, more delicately and temporarily only, by local cooling, electrical polarization, by infusion of local anesthetics, by application of KCl and consequent spreading depression, or by placing a local, nonspreading seizure. With the oncoming of 'wet' neurophysiology, the 'new science' of the so multifaceted and complex neurotransmitter interfaces and the neurotransmitter pathways, more discrete and specific techniques for making local lesions became available. These humoral transmitting mechanisms can be reduced in their activity ('conductance'), if not completely silenced, not only by classical highly localized electrolytic lesions, but also and more specifically by local or intraventricular application of neurotoxins (e.g. 6-OHDA for catecholaminergic transmission mechanisms). 'Lesions' of a more transitory and often 'milder' nature can be engendered by such pharmacological tools as specific (e.g.

---

[1] We use the term 'behavior' in the present context quite broadly. It refers to 'exteroversive' – overt – motor and/or autonomic activities (observable from the outside) as well as those 'internal' – covert – activities such as thinking, feeling, and remembering, observable either through our conscious mind or then indirectly by consequential 'exteroversive' (e.g. verbal) behavioral patterns. This definition does not clash with *MacLean's* [71] notion that both the psyche and (exteroversive) behavior are manifestations of information.

postsynaptic) receptor blockers, or – acting on the presynaptic side of the interneuronal gap – by synthesis and release inhibitors.

As to *stimulation,* one can again use the 'classical' method. Through adequately placed electrodes one can apply an electrical current of proper polarity, waveform, intensity, and frequency to produce a focus of enhanced local activity. As with the lesion technique one observes, usually in freely moving, unrestrained animals, changes in particular behaviors or, possibly, the appearance of a 'new' behavior. The 'wet' counterpart of stimulation can be seen in an enhancement of activity in specific transmitter systems. This is achieved by intracerebral application of the transmitter itself, by a local or systemic 'load' with precursor substances (e.g. *L*-DOPA to drive dopamine and/or noradrenaline transmission activity), by facilitating release (e.g., by a block of presynaptic α_2-receptors or by 'releasing' substances such as amphetamine or tyramine), by inhibiting reuptake, by slowing down metabolic inactivation of the transmitter substance, or by facilitating interaction of the humoral information carriers with the receptors (e.g., using diazepam to enhance GABAergic transmission). In general, these 'wet' stimulation techniques are characterized by a relatively high specificity as concerns selected transmitter systems and receiving areas. Both, the 'wet' lesioning and stimulation techniques belong to the experimental armamentarium of the neuropsychopharmacologist.

Recording in the classical way is done mainly by electrophysiological techniques. One monitors by properly placed electrodes the 'spontaneous' (local) electroencephalogram in its particular frequency/power characteristics, the activity patterns of single units, evoked responses or 'slow' (i.e. basically DC) electrical phenomena. In general, electrical recordings are performed during ongoing behavioral observations. They indicate the qualitative and quantitative involvement of the recorded structure in the organization of the behavior or behavioral component under investigation. Recordings can also be done with 'wet' methods; they give indications on the involvement of particular transmission systems in a given behavioral activity. Usually, while an interesting behavior is in progress (or has been induced by environmental manipulations), the animals are sacrificed and transmitter concentration and/or turnover are measured in various brain areas. Often push-pull cannulas are implanted to measure transmitter release in a predetermined site. At present, there are efforts to develop methods that allow monitoring local transmitter activity or release by implantable 'chemtrodes'. Measurement of ligand binding has been introduced to supplement 'pure' transmitter activity determinations by an indicator of receptor sensitivity.

In general, all three neuropsychophysiological approaches – lesion, stimulation, and recording – aim at the establishment of structure/function relations; one attempts to indentify those (usually multiple) CNS structures, systems or networks that are intimately involved in the organization and regulation of a particular behavior. One searches for the central neural and neuronal, the physico-chemical, organizational basis of behavior.

We intend, in the present paper, to identify and investigate in particular those functions whose malperformance gives rise to the symptoms that characterize the afflictions under discussion. It is obvious that, when working on animals, any functional derangement and its symptomatic manifestations must always be seen in connection with the behavioral repertoire of a given species. Whatever these manifestations mean in another species is, at this time, still a matter of speculation or, at best, of an educated guess. With these methodological and conceptual guidelines, restraints, as well as liberties in mind, we shall present in the following section some older and newer experimental data that are liable to offer the information as defined above. The reader interested in more detailed information on the phenomena reported here is referred to the papers cited in the text and, in particular, to the outstanding writings of MacLean [69–71] or to the volume of *Adey and Tokizane* on the 'Structure and Function of the Limbic System' of the Progress in Brain Research Series.

Evidence on Limbic Functions from Animal Experimentation

The Klüver-Bucy Syndrome

It is not by accident that we begin this section with a discussion of the classical experiment of *Klüver and Bucy* [58]. Almost overnight that study converted our till then rather faulty notions about the 'olfactory brain' – except perhaps for *Papez'* [96] theory – into some good and still valid ideas about the functional roles of (at least some aspects of) the limbic system.

Klüver and Bucy [58] placed large bilateral lesions in the temporal lobes of rhesus monkeys. The lesions involved the uncus, the amygdala and parts of the hippocampus. These lesions produced a set of behavioral changes consisting of tameness (i.e. loss of fear and aggressivity), hypersexuality, excessive oral exploration, changes in dietary habits, visual agnosia and excessive reaction to visual stimuli. Without going into a more detailed description of the symptomatology of these operated animals [for this one may consult refs. 10, 56 – 58], one can deduce from the 'cardinal symptoms' that obviously intactness of the anterior portion of the temporal lobe, parts of the hippocampus, the hippocampal gyrus, and the amygdalar complex including the many projection areas of these structures, is a necessary condition for proper organization and control of mood (fear and/or aggression), sexual drive and behavior, oral exploratory activity (as a component of general exploratory activity), visual perception (hypermetamorphosis), food selection and, possibly, habituation. Many investigators have attempted, and in part succeeded, to confirm the results obtained by *Klüver and Bucy* and, through more discrete lesions, to analyze further the structure-function relation of single components of the syndrome.

Aggression, Fear, Tameness

As to 'tameness', *Schreiner and Kling* [109, 110], destroyed bilaterally the amygdalar complex in a mountain lion and other felines and thus produced 'tame cats'. Here it is of interest to note that animals initially made tame through amygdalectomy were turned into savage beasts again by additional lesions in the ventrolateral area of the hypothalamus.

From these first experiments one may conclude that the amygdala per se exerts a facilitatory influence on aggressive behavior. However, closer scrutiny with additional, still more discrete, experimental approaches quite clearly showed that the situation is more complex by far; that the amygdala can exert an enhancing or an attenuating influence on aggression; it was noted also that this structure is involved in the control of other behaviors; and that other areas are also involved in the organization and control of fear and aggressive behavior. *Wood* [125] noted that electrical stimulation in the lateral amygdaloid nucleus of the cat failed to produce emotional behavior, whereas stimulation of the basal nucleus was followed by responses signalling fear and/or anger. Bilateral lesions in the central amygdalar nucleus resulted in increased aggressive behavior as well as increased food intake. In turn, stimulation of these structures led to fear and anger responses. Finally, lesioning of the medial nuclei produced hyperactivity.

At this point it is of interest to recall that rage can also be produced by electrical stimulation of the perifornical area of the diencephalon of the cat [33]. *Hunsperger* [36] has been able, through electrical stimulation of various structures, to delinate a neural system, extending from the midbrain up to the limbic brain of the cat, which seems to be involved in the initiation and elaboration of emotional behavior, such as rage, aggression, fear and flight.

Egger and Flynn [19] found that electrical stimulation of the basal amygdala and of the anterior and medial portion of the lateral amygdalar nucleus suppresses hypothalamically elicited attack behavior in cats, whereas lesions in this area facilitate this behavior. In turn, stimulation of the dorsolateral part of the posterior-lateral nucleus of the amygdala facilitated attack behavior.

Hypothalamically induced attack behavior can be inhibited by simultaneous electrical stimulation of the ipsilateral, but not of the contralateral, lateral prefrontal cortex [111]. *Nakao* [8] has shown that cats, after lesions were placed bilaterally in the amygdala, became less responsive to hypothalamic electrical stimulation and reacted less vigorously to dogs.

After ablation of the cingulate gyrus Macaca mulatta shows increased aggressiveness (or more 'fearlessness') in response to man; an effect which disappears, however, in the course of 2 months [78]. Hippocampectomized rats (50–70% removed with only the anterodorsal and ventral parts remaining) reveal 'less fear' in an avoidance task; they are impaired in their association of a 'neutral stimulus with the responses elicited by noxious stimulation' [93].

According to *Green* et al. [30] cats become 'placid' after amygdaloid lesions but tend to develop rage when the lesion also involves the hippocampus (with ensuing seizures). Stimulation of the amygdala of monkeys also induces, according to *Anand and Dua* [3], fear and rage. Rats can be made aggressive by septal lesions [52] as well as by ablation of the olfactory bulb [119]. In our own [unpublished] work we noted that electrical stimulation of the dorsal hippocampus of cats has little influence on the rage produced by hypothalamic-perifornical stimulation. Amygdaloid lesions that relieve hypothalamically induced rage have little effect on the rage engendered by septal lesions [53]. According to *Adamec* [1] medial and ventromedial hypothalamic stimulation in cats stops the spontaneous predatory attack.

Kling et al. [55] recorded and computer-analyzed the (telemetrically transmitted) EEG of the amygdala of African Green Monkeys during social interaction and when alone. They noted the highest power (mainly in the theta band) during passive approach, yawning ('passive threat'), passive genital inspection and run and chase. Low power output was observed during active and passive grooming, sitting together huddling, sitting alone and during sleep. The authors interpret their findings to indicate that amygdalar activity is related to 'emotional significance and degree of ambiguity of an interaction'. According to *Rosvold* et al. [103], amygdalectomy leads to a drop in social dominance in monkeys. *Jonason* et al. [40] in their detailed work in rats placed successive septal-amygdalar or amygdalar-septal lesions. When the septal lesion was inflicted first, social 'cohesiveness' increased and was then decreased again following the amygdalar lesion. In turn, an initial amygdalar lesion reduced cohesiveness which was brought back to control levels after the septum was lesioned.

In the work of Gotsick et al. [26] rats were initially exposed either to fighting, to apomorphine treatment, or fighting together with apomorphine treatment. Only the animals subjected to the combined (behavioral and drug) experience showed increased fighting activity in response to (later) apomorphine administration. This suggests not only the involvement of dopaminergic pathways in the organization of this type of aggressive behavior, but also some sort of 'sensitization' when fighting and enhanced dopaminergic receptor activity occurred together in the 'history' of the animals.

As to transmitter participation it is also of interest to mention the work of *Daruna and Kent* [14] on intraspecies aggression. They noted enhanced serotonin turnover rates in the anterior forebrain and the brainstem of 'high aggressive' rats in comparison with 'low aggressive' animals. *McLain and Cole* [76] demonstrated that pretreatment of rats with para-chlorophenylalanine (PCPA), a serotonin-depleting agent, decreased the latency of 'frog attack' but had negligible effect on its frequency. PCPA and alpha-methyl tyrosine (a catecholamine depletor) pretreatment increased the frequency of 'mouse attack'. In the work of *Mark* et al. [74] 'rage' cats were produced by ventromedial hypothalamotomy. Injection of minute amounts (up to 0.18 mg) of epinephrine or norepinephrine bilaterally into amygdala had a pronounced taming effect on these animals.

The suppressant effect of punishment on operant behavior – interpretable as a manifestation of fear or anxiety – is reduced (i.e. 'less fear') following local injection of norepinephrine into the amygdala of male rats. Local pretreatment with an α-adrenergic blocking agent (phentolamine) but not with a β-adrenergic blocker (LB-46) counteracts this 'antianxiety' effect of norepinephrine [73]. Beta-agonists, however, facilitate the suppressing effect of punishment. *Jaekel* [in preparation] has shown that in monkeys and rats β-adrenergic antagonists quite effectively reduce 'fear'; where and how, however, is not known.

It is of interest to learn – although a direct link to limbic function is not established – that histamine given intraventricularly dampens the aggressivity induced by electrical foot shocks in rats, whereas 4-methylhistamine facilitates aggression [101]. The authors conclude that 'H_1-receptors are inhibitory and H_2-receptors facilitatory in footshock aggression.'

Sexual Behavior

One of the outstanding symptoms of the Klüver-Bucy syndrome is hypersexuality, suggesting that the lesion in those monkeys had affected one (or several) structure(s) that are involved in the control of sexual behavior and drive. *Wood* [125] had noted that hypersexuality (in male cats) is produced by lesions restricted to the lateral amygdaloid area. Similar observations were made by a number of other authors, e.g. *Kling and Schwartz* [54]. In turn, *Green* et al. [30] had found that removal of the piriform cortex in cats is sufficient to induce changes in sexual behavior. The study of *Hunsperger* [37] suggests enhanced sexual 'drive' after circumscribed lesions in the amygdala and the piriform cortex; mounting frequency is increased and 'neck grip' intensified.

According to *Kling* et al. [53] hypersexuality induced in male cats by amygdalar lesions is abolished by septal lesions, suggesting that this latter structure is involved in the (positive?) control of sexual behavior. This is supported by the observation of *MacLean and Ploog* [72], who were able to elicit in the squirrel monkey penile erection through electrical stimulation in the septal area. This same structure seems to play a role in the sexual behavior of female animals, too; *Sawyer* [108] noted electrographic signs of activation in the septal area of female rabbits coincident with orgasmic behavior.

The importance of the septal area for sexual behavior is further evidenced by sex-linked morphological differences. *Raisman and Field* [100] had found that synaptic endings in male rats differed from those in females. More recently *Nishizuka and Arai* [87], reporting on an electron microscopic study, noted that in the medial amygdaloid nucleus the number of so-called shaft synapses, spine synapses and somatic synapses per surface unit was considerably higher in male rats as compared to females. This again suggests a sexual dimorphism, here in the amygdala, and thus points to this structure as an important area for the organization of sexual behavior.

In this connection it is of interest to mention the study of *Kawakami* et al. [45], who noted that through electrical stimulation of the amygdala *and* the dorsal hippocampus one can modulate the serum concentration of LH and FSH. *Kawakami and Kimura* [44] found a clear influence of estradiol treatment on the amplitude of the hippocampal and amygdalar EEG phenomena.

'Oral' Tendencies

The monkeys of *Klüver and Bucy* [58] showed clear signs of exaggerated oral exploration. They examined objects, even inedible ones, directly by mouth instead of picking them up first with their hands. They even used the 'direct oral approach' on such live objects as snakes, mice or cats. *Hess* [33] noted in the course of his stimulation experiments such oral tendencies as licking, chewing, salivation, but also 'forced' eating (bulimia) from stimulation sites in the preoptic, septal and anterior hypothalamic region.

Of interest in this connection are the observations of *Wood* [125], who upon electrical stimulation in the basal amygdaloid nucleus noted his cats to bite themselves, or the floor, while showing 'signs of anger' and running in circles. Stimulation of the external capsule yielded 'oral responses' (licking, chewing, swallowing, salivation). Similar observations were made by *Kaada* [42] and *Ursin and Kaada* [118]. The exaggerated neck grip of male cats already mentioned [37] may also be cited as an 'oral sign' (here in connection with sexual behavior). However, these oral tendencies may also relate to functions involved with hunger, food intake and/or satiation. In fact, *Green* et al. [30], *Fuller* et al. [23], and *Fonberg and Delgado* [22] suggest a role of the amygdala as a controlling instrument of the well-established 'hunger and satiation centers' in the hypothalamus. *Lewinska* [64] placed lesions either in the basal parvocellular part, or the cortical nuclear part of the amygdala of cats. In the first case she noted increased, in the second case decreased food and milk intake. Electrical stimulation of the lateral hypothalamus, the posterior hypothalamus, or the amygdala is followed by an increase in gastric secretion. In turn, stimulation of the ventral medial hypothalamus and of some extraamygdalar structures of the limbic system results in a decrease of secretion. Lesions in these areas are followed by effects contrary to those produced by stimulation [9, 31]. Polypeptides such as TRH [83], β-endorphine [107], or bombesin [117] may also be involved; GABA concentrations (in the hypothalamus) and gastric secretion increase during hypoglycemia [51].

Self-Stimulation, Reward, Pleasure, Motivation

Olds and Milner [91] observed that rats given the opportunity to turn on the stimulating current to excite certain of their own brain structures seemed to experience 'pleasure' and learned to push the lever with little indication of fatigue for extended periods of time. To experience this 'self-induced' pleasure they were found to be ready to forfeit food and water, to forget about sex and even to accept pain (i.e. an electrified grid in the path to the lever). In the first experimental series it was noted that the stimulating electrodes had to be in or in the neighborhood of limbic structures – in particular the medial forebrain bundle, the septal area and the amygdala – to produce this positive reinforcement effect. *Olds and Milner* [91] in fact initially thought that this reinforcement paradigm involved *Nauta's* [85] mesencephalic-limbic system with its rostrally projecting pathways. Additional indication for the participation of limbic structures in intracranial self-stimulation stems from *Ito and Olds* [38]. Self-stimulation periods were found to be attended by increased unit firing in the cingular region and inhibition of discharge in the hippocampus.

Crow [12, 13], *Routtenberg* [104], *Routtenberg and Malsbury* [106], *German and Bowden* [24], and *Wise* [123] made a strong point for the participation of 'limbo-petal' catecholaminergic pathways to be relevantly involved in intracranial self-stimulation. *Stein and Wise* [113] and *Arbuthnott* et al. [4] noted enhanced noradrenaline release in such areas as the amygdala and the preoptic area during self-stimulation. *Wise* [122], in a recent article, critically reviewed the many pieces of evidence that support the notion of catecholamine participation in self-stimulation. *Stein and Belluzzi* [112], collecting the already substantial evidence, discussed the possibility that (limbic) opiate systems may also be involved in intracranial self-stimulation.

Hoebel and Teitelbaum [35] had some evidence for an obvious functional *and* topographical relation between feeding *and* intracranial self-stimulation. In a review article *Hoebel* [34] again pointed out the importance of this interrelationship. Damage to the hypothalamic 'satiety system' or to the ascending catecholaminergic system in the lateral hypothalamus can disinhibit self-stimulation *and* feeding.

Routtenberg [105] suggested that the 'brain stimulation reward system may function as a memory consolidation system'. He put forward the hypothesis that ascending dopaminergic pathways participate in the memory consolidation process, possibly by facilitating protein phosphorylation in forebrain structures, providing a 'biochemical residual following learning'. *Mondadori and Waser* [79] also make a strong point in relating reward to memory functions.

According to *Goddard* [25] the amygdala is involved in the suppression of motivated approach behavior. Amygdalectomized animals have difficulties in associating fear and avoidance behavior with previous neutral stimuli. Along with this view, *Jacobs and McGinty* [39] noted unit activity in the basolateral amygdala to be positively correlated with behavioral inhibition and to be 'tuned' to respond to complex environmental stimuli. *Ursin and Kaada* [118] on the basis of their stimulation experiments suggested that the amygdala plays an important role for behavioral attention. *Lopes da Silva and Kamp* [67] observed in the hippocampus of dogs a temporary increase in the 5–6/s theta activity when the animals had to press a lever for reward in response to light or sound stimulation. Reviewing a number of 'relevant' neuroanatomical and electrophysiological findings, *Kimble* [49] came to the conclusion that the hippocampus has an inhibitory function; that 'it constitutes part of the neural machinery necessary for the generation of a brain process which is functionally equivalent to Pavlovian inhibition'.

Limbic System and Memory

A considerable amount of the earlier work already suggested that parts of the limbic system, particularly the hippocampus, participate in those complex mechanisms which control storage and/or retrieval of information. Defects in memory, in general comparable – mutatis mutandis – to those described in clinical cases, have been observed after resection of the medial temporal lobe in monkeys by *Orbach* et al. [95]. *Grastyan and Karmos* [27] lesioned the hippocampus bilaterally in cats and noted a marked impairment in the delayed reaction. *Kaada* et al. [43] observed slowing in maze learning of rats after bilateral hippocampectomy.

In connection with these earlier lesion experiments it is of interest to learn about changes in electrical activity of the hippocampal area associated with conditioning. *Grastyan* et al. [28] found that the 4–7/s hippocampal waves of the cat were more pronounced in the early stages and that desynchronized patterns attended later stages of conditioning. There is pronounced theta activity in the hippocampus during the (appetitive?) approach phase of conditioned responses [2]. *Olds and Hirano* [90] observed marked increments in both the rate of firing and number of responding units in the hippocampus (and the thalamus) of rats in response to (conditioned) auditory stimulation.

Niki [86] bilaterally removed by aspiration portions of the hippocampus of rats. On the basis of his results he made the following statements: Hippocampal ablation results in disinhibition of previously extinguished bar pressing; it increases the rate of responding to nonreinforced stimuli while not affecting the response to reinforced stimuli; it impairs position reversal learning; it produces a severe deficit in single alternation learning; it reduces flexibility of solutions in a Dashiell maze; it enhances perseverance in maze behavior; it slows down habituation of the cardiac response to sound stimuli. According to *Niki* [86] the hippocampus plays an important role in the inhibitory control of behavior. *Poschel* [99] found that through hippocampal stimulation – and thus, induction of a hippocampal seizure – the rat's learning abilities and (long-term) retrieval functions are blocked – in this author's opinion a Korsakoffian type of memory deficit. *Kimble and Jacobson-Kimble* [50] investigated the influence of bilateral hippocampal lesions in rats upon the behavior during acquisition and extinction in a brightness discrimination task. The authors noted that in the lesioned rats there was no increase in number of trials necessary to reach criterion but that such animals generated a significantly longer 'position hypothesis' during acquisition.

Olton [92] and *Olton* et al. [94] tested rats in a 'new spatial discrimination procedure'. They found that destruction of the entorhinal area, the fimbria-fornix system anterior to the hippocampus, the septum, or the postcommissural fornix produced a marked and consistent impairment in performance. The authors concluded that the hippocampus (with its various afferent systems) plays an 'important role in processing information about spatial location'. *O'Keefe and Dostrovsky* [89] and *Black* et al. [7] from their experiments with hippocampal lesions, concluded 'that animals with such lesions cannot process information about places and therefore cannot employ place strategies in avoidance learning and punishment situations.' *O'Keefe and Black* [88] more recently, recording single cell activity in response to sensory inputs to the hippocampus, produced additional evidence supporting the hippocampal cognitive map theory. Also *Olton* [92], with additional data obtained with the lesion and stimulation techniques, was able to support his notion about the importance of the septo-hippocampal system for 'spatial behavior'.

Gray et al. [29] studied the effect of various lesions in the septo-hippocampal system upon extinction in an alley after continuous or partial reinforcement. In their view the septo-hippocampal complex constitutes a 'behavioral inhibitory system', inhibitory though in terms of the stimuli it responds to (conditioned punishing stimuli, conditioned frustrative stimuli and novelty). *Gray* et al. [29] also found connecting lines between their notion and *Weiskrantz and Warrington's* [121] 'revolutionary' view on amnesia: 'an amnesic subject is amnesic when prior learning is allowed to interfere with subsequent learning and scarcely amnesic at all when false positive responses are not allowed to intrude', as cited by *Gray* et al. [29]; and '... the human amnesic's problem lies in inhibiting responses which under earlier conditions were correct but no longer are so'. *Landfield* [63] investigated in rats the effect of driving or blocking the hippocampal theta rhythm (by low and high frequency, respectively, electrical stimulation in the septum). He noted that theta driving improved retention performance in a one-way active avoidance task in comparison with animals receiving high frequency stimulation or nonstimulated controls. In a one-way passive inhibitory task 'driven' animals performed better than the 'blocked' ones. Low frequency stimulation during testing also improved performance in the active avoidance task. *Landfield* [63] suggests his results to indicate that the theta rhythm is associated with memory storage processes. In the present writer's view [59] the theta rhythm could well be the neural (e.g. hippocampal) manifestation of a high level of 'Bereitschaft' or local vigilance; a level necessary to adequately perform and attend to the task at hand.

Rolls [102], using the microelectrode technique, found neurons in the monkey's extreme anterior thalamic area to respond to visual stimuli only when these were familiar. 'Recognition memory' seems to involve this particular diencephalic area (having close connections to the limbic system). These experimental results may well constitute a connecting piece of information towards a better understanding of the 'visual agnosia' noted in the Klüver-Bucy syndrome. In this connection it is of interest to learn that septal area stimulation in cats attenuates photically evoked potentials as recorded from the lateral gyrus [68].

Krieckhaus et al. [62] lesioned the septum in Long-Evans rats. All lesioned animals, including those that did not show the (otherwise characteristic) hyperemotionality syndrome after the lesion, performed better in a conditioned avoidance response paradigm. According to the authors the 'septal syndrome' is not a necessary precondition for improved learning and 'septal lesions (evidently) weaken fear-elicited unconditioned responses, which usually compete with the instrumental avoidance response.'

In rats subjected to hippocampal or cortical lesions and in unoperated controls, *Means* et al. [77] investigated maze activity, spontaneous alternation, food-reinforced alternation, preservation of the alternation response and the response to a novel alley before and after training. They noted that the hippocampally lesioned animals were more active in the maze and deficient on spontaneous and food-reinforced alternation as compared with cortically lesioned and control rats. Animals with hippocampal lesions did not perseverate in the alternation response and did not enter a novel alley after previous training.

Dreyfuss and Mühlethaler [17] have shown that various vasopressins given locally to hippocampal slices were able to modulate the firing of (presumptive) pyramidal cells. These findings relate quite clearly to *de Wied's* [16] discovery that vasopressin affects memory and as a well to *Buijs'* [11] results suggesting the existence of 'vasopressinergic' fibers situated in the ventral hippocampus and terminating at dendrites of pyramidal cells.

Sternberg and Gold [115] have been able to show that subseizure electrical stimulation of the amygdala of rats (without regard to more specific localizations) produced signs of retrograde amnesia. In view of specific neurotransmitter inputs, it is of interest to learn that (both α- and β-) adrenergic antagonists attenuated amnesia produced by this limbic stimulation for visual discriminatory training. Evidently, adrenergic mechanisms are involved in the (still not quite clearly understood) mechanisms leading to amnesia.

Stereotypies

One may question the relevance of this artificially (usually pharmacologically) induced symptom in rats within the framework of a review on limbic functions. Still, as this 'model' – the amphetamine- or apomorphine-induced stereotyped behavior – is commonly used as a screening test for neuroleptic drugs, a few points may be mentioned here. Earlier interpretations saw drug-induced stereotyped behavior as the manifestation of enhanced activity in the nigrostriatal and the mesolimbic dopaminergic systems. However, *Kelly* et al. [46] had shown with 6-OHDA-induced local lesions that the intact striatum seems to be of more importance for the maintenance of this symptom than is the accumbens nucleus, an observation that should keep us from indiscriminately assuming that the accumbens nucleus is a candidate for 'stereotypogenic' activity. Still, this nucleus together with other limbic structures in that area may well be intimately involved in the organization of 'normal' behavior. According to *McKenzie*

[75] bilateral ablation of the tuberculum olfactorium markedly reduces stereotypies induced by apomorphine in rats. In turn, bilateral severance of the olfactory bulb or large bilateral lesions in the striatum had little influence on the apomorphine-induced 'dyskinesia', an observation which is again in contrast to the work of *Kelly* et al. [46]. *Stevens* [116] discussed the possibility that the accumbens (and the striatum) may exert an important 'gating' influence if it comes to the integration of affective, ideational and visceral influences in the making of 'normal' behavior. In turn, *Grinberg-Zylberbaum* et al. [32] had shown that rats, when stimulated in the striatum are prevented from acquisition of instrumental behavior.

Sleep, Waking, Vigilance, Attention

Parmeggiani [97] has shown that a sleep-like state or at least 'presomnic' signs can be induced by low frequency stimulation of the hippocampus and the fornix. Theta waves appear in the hippocampus with the onset of paradoxical (i.e. low voltage) sleep [41, 61], an electrographic pattern which is otherwise seen when animals are aroused by novel stimuli. *Morales* et al. [82] investigated the interrelation between the hippocampal theta activity and unit discharge bursts of the septal area in the rat. They found that during paradoxical sleep as well as during restless wakefulness there was a striking correlation between these two electrical patterns. In turn, during quiet waking and slow wave sleep the hippocampal theta activity and the septal burst discharge disappeared.

The appearance of theta waves in the hippocampus during arousal – in particular during orienting [47] – and during paradoxical sleep is of paramount importance as this phenomenon somehow seems to signal a state of high local vigilance or attention; to search the surroundings in the one case, to follow the scenery of the dream (if animals dream?) in the other [59]. In the experiments of *Black and Young* [8] dogs were trained to 'hold still' or press a lever in response to (discriminatory) stimuli. The authors noted that under 'move' conditions the hippocampal theta activity was more pronounced and of longer duration than under 'hold still' conditions. In her classical paper, *Kennard* [48] after lesioning the cingulate gyrus in cats and monkeys noted clear signs of 'altered consciousness'. The cats, after recovery from surgery behaved as if they were 'less aware of the surroundings', they ingored or failed to recognize food, they were slow and hypokinetic, they bumped against obstacles and suffered from exaggerated perseverative behavior but had no signs of paresis or of postural abnormalities. Similar symptoms were observed in monkeys which in addition were 'soft' and 'uncompetitive'.

Unit responses and the cortical EEG were recorded in the dorsal hippocampus of unanesthetized rabbits in the study of *Lidsky* et al. [66]. They found that the unit discharge rate was not linearly related to arousal level.

Autonomic Activity

We can restrict our discussion with respect to autonomic indicators of the limbic function to some classical experiments. *Kaada* [42] in his extensive work in the limbic forebrain observed manifold effects in the autonomic sphere in response to stimulation of these structures, such as changes in blood pressure, heart rate, respiration, intestinal activity, pupillary size and piloerection. It is probably not too far-fetched, if one sees these autonomic signs as vegetative concomitants of the various behavioral activities induced by limbic stimulation as described earlier. Still, it is not unlikely as well that these autonomic stimulation effects may be the manifestations of a true vegetative regulatory role of these limbic structures, superimposed over the 'classical' head ganglion for autonomic regulation, the hypothalamus [33]. In this connection one should make reference to *Hess'* [33] well-known work on the diencephalon and adjacent areas. He noted, in response to electrical stimulation in the preoptic and septal areas of the cat circulatory and respiratory effects as well as panting. A variety of 'oral' phenomena were mentioned earlier. Of particular interest in the present connection are such autonomic effects as defecation and micturition in adequate posture, a whole 'package' obtained in cats upon electrical stimulation of the septal area. Such compound effects certainly point to the 'integrative power' of the limbic system.

Neurotransmitter Aspects

We had mentioned on several occasions that in the neural organization of the different behavioral patterns of interest in the present context, various neurotransmitters play an important role. This cannot surprise, as the large majority of neuronal networks involved in the control of behavior *has* to rely for interneuronal information exchange upon 'wet' carriers; the various structures and the intrinsic and connecting networks of the limbic system cannot be an exception. In fact, since the early days of the 'neurotransmitter science' we have ample evidence that aminergic, aminoacidergic, *and* polypeptidergic neurons, project to, and make connections within and between the various rhinencephalic nuclear and cortical areas. The pertinent question thus is rather, as to what neurotransmitter systems are involved in what function and in what area for the neuronal control of these behaviors. At this time we can state that a multitude of different transmitters *must* partake in the organization of even the simplest behavioral activities; a 'one behavior/one transmitter' relation does not exist [60]!

The evidence presented so far in this chapter suggests that in the organization of the behavioral symptoms signalling aggression, fear, tameness, and predatory attack patterns, dopaminergic, α- and β- adrenergic, and serotonergic transmission mechanisms are involved. The participation of sex hormones, apart from other transmitters, as modulators of neuronal activity in the control of sexual behavior is also quite obvious. There is evidence that adrenergic, GABAergic and polypeptidergic transmission lines partake in the organization of hunger and satiation and thus, quite probably, in that of 'oral behavior'. The reinforcement effect of self-stimulation – thus reward and possibly motivation and pleasure – seem to rely on intact adrenergic, dopaminergic, polypeptidergic and certainly on GABAergic links. For the quite complex business of 'memory', there is good evidence that vasopressin, ACTH fragments, but also α- and β-adrenergic, dopaminergic, serotonergic, GABAergic and cholinergic channels [15] play a pertinent role.

The multiplicity of (qualitatively different) wet transmission mechanisms involved in the control of sleep and waking – serotonin, noradrenalin, dopamine, glycine, GABA, acetylcholine, delta sleep-inducing peptide, enkephalin, vasotocin, etc. – was recently reviewed by *Koella* [60]. Stereotypies – or the absence of stereotypies, i.e. 'normal' behavior – do depend on particular levels of activities in dopaminergic, and, for feedback, in GABAergic and polypeptidergic channels.

Some recent evidence of 'limbic biochemistry' corroborates and refines (the mechanisms involved in) these 'wet' links in the neuronal organizational patterns. *Moore and Bloom* [80, 81], largely drawing on their own work, have reviewed the anatomy and the cellular physiology of the dopaminergic, noradrenergic, and adrenergic projections to the limbic system and the many other central nervous receiving areas. *Bischoff* et al. [6], using the spiperone binding technique, demonstrated the presence of dopamine receptors in the rat hippocampal formation. *Epelbaum* et al. [20] showed that noradrenaline stimulates the release of somatostatin – a polypeptide – from slices derived from the rat amygdala and hippocampus; with these results they reactivated our interest in the multiplicity of patterns of interaction between the various transmitter systems (which does not make the understanding of neuronal organization any easier!) In their work in the rat, guinea pig, and bovine brain, *Peroutka and Snyder* [98] demonstrated the existence of two distinctly different serotonin receptors in the hippocampus (and other brain areas).

Continuing the classical work of *Lewis and Shute* [56], *Benardo and Prince* [5], in experiments on hippocampal slices, demonstrated the intimate mode and mechanisms of action of acetylcholine on the pyramidal cells of this structure. *Flood* et al. [21] using a variety of cholinergic agonists and antagonists, obtained information about the 'proper' cholinergic activity necessary for adequate retention performance of mice.

New evidence about the local effects of benzodiazepines and barbiturates on CA1 pyramidal neurons of the rat's hippocampus was used by *Wolf and Haas* [124] to support the present view that these drugs act via facilitation of GABAergic transmission.

Duka et al. [18] produced evidence for a multiplicity of opiate receptors in the hippocampus of rats. The presence of both, enkephalin and zinc in the rat's hippocampal mossy fiber system was demonstrated by *Stengaard-Pedersen* et al. [114]. *Zieglgänsberger* et al. [126] were able to excite, in the anesthetized rat, hippocampal pyramidal cells by various (natural) opiates. They interpret their findings to be the manifestation of disinhibition; i.e. inhibition of inhibitory interneurons. And finally, *Vincent and McGeer* [120] produced evidence for a substance P input to the rat's hippocampus from the septum, running in parallel to a cholinergic pathway.

Limbic Seizures

One could make a long story short and just show a picture indicating the time course, intensity, and nonspread of hippocampal seizures induced by short electrical stimulation of this very same structure (fig. 1). However, *MacLean's* [69, pp. 36, 37] description of this phenomenology is so impressive and so telling in connection with the topic of this symposium that we prefer to cite it verbatim:

'More dramatic changes are seen during afterdischarges which follow stimulations of greater strength. Any turning movements that were present during the initial stimulation appear to reverse themselves with the onset of the afterdischarge, and the pupils may suddenly dilate. Purring, if previously present, may cease and be replaced by occasional meows or yowls. Concurrently, the animal assumes attitudes that strike one at first as being rapt atten-

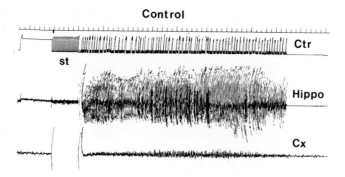

Fig. 1. Limbic seizure induced in the freely moving cat, prepared chronically with recording and stimulating electrodes, by a short train of 5 shocks per second (see left side in 'counter write-out') to the left hippocampus. Note that seizure begins already during stimulation but is not recorded until cessation of the stimulus, and lasts for some 40 s. No spread to cortex (Cx). Ctr = Automatic spike counter. Note abrupt cessation of discharge. The (freely moving) animal shows no overt signs of seizures except occasional limb jerks. During seizure it 'stares' into space and is usually motionless except perhaps for some adjustment of posture. After cessation it usually vocalizes in a 'complaining' way and performs some 'orienting' or searching movements.

tion or fearful alerting for the unexpected. Further examination indicates, however, that the animal is poorly in contact with his environment. Although the pupils react to light, the animal will not avoid the light, nor will he cringe when one pretends to strike the face. If one forcefully blows smoke at him, he will withdraw a little, but will not avoid the smoke. A burning cigarette may be brought up to his nose with no response, but if there is momentary contact with the lighted end, the animal may lunge about wildly. The emotional changes that one may induce during an afterdischarge are most interesting: if one takes an animal which prior to stimulation simply meows when the tail is pinched, one finds that the same noxious stimulus will cause it to hiss, spit, and strike out during the afterdischarge. A prolonged or repeated noxious stimulus may lead to states of wild excitement, accompanied by spitting and hissing. I have held on to the tip of the tail of such animal and observed a state of rage in which he would not properly orient his attack, but rather would viciously bite at the stump of his tail. It is striking that these excited states may suddenly terminate in rather prolonged catatonic-like stances. A meow or series of meows usually signals the end of the afterdischarge.'

As to spread of these discharges *Kaada* [42] states that afterdischarges initiated in the hippocampus readily spread into the limbic cortex of both sides and further to the cortex just lateral to the rhinal fissure, this even in anesthetized cats and dogs. Based on his own observations in drug-immobilized cats, squirrel monkeys, and macaques *MacLean* [69] states: 'it is to be emphasized that there may be a propagation of hippocampal discharges without a discernible alteration of neocortical activity.'

Some Interpretative Thoughts

This short review on the psychoneurophysiology of the limbic system was not intended to be complete or close to complete. We rather chose to select a number of experimental results – not rarely of older vintage – that somehow should characterize, mainly in principle, the behavioral (and in part neural) effects obtained by manipulating the activity in various parts of these areas of the CNS. These data were supplemented by some references to recording experiments.

A few important points seem to emerge: first, from both the lesion and stimulation experiments, i.e. from the 'symptomatology' of the effects produced by locally depressing or enhancing activity in some parts of the limbic brain, one has to draw the conclusion that invariably complex functions are affected. Unlike stimulation (or lesion) foci in the cerebral motor cortex that affect in a point-to-point fashion single muscle groups or even single muscles, the interventions in the limbic system affect whole behavior patterns, i.e. activities that involve an immense variety of effector organs and thus must be based on a complex central regulatory and controlling system of integrating networks.

Secondly, many but not all of the behavioral symptoms elicited by limbic manipulations quite clearly bear some relation to the 'psyche', or at least to what may be interpreted to be the psyche in animals. The symptoms seem to signal changes in mood (e.g. anger and pleasure), drive, motivation, attention, but also in the multifaceted memory functions and the sexual, recognition and orienting behavior. Other effects of limbic manipulations, e.g. micturition or defecation in adequate posture, rather may be taken as manifestations of 'lower', though highly complex, functions.

Thirdly, it is difficult to assign a priori one function to one particular component of the limbic system. Effects manifesting changes, for instance, in mood or drive or memory functions can be elicited by stimulation or lesions in a variety of limbic structures. The functional interrelations between the various parts of the limbic brain are well documented by experiments in which behavioral changes induced by intervention in one part can be modulated – positively, negatively and/or qualitatively – by stimulation or focal lesions in other parts. For most of the behaviors of interest in the present context, one gains the impression that the limbic system handles its organizational tasks as a whole, and quite certainly in functional interaction with lower and higher structures. This should not mean to indicate that the various components of the limbic brain, due to their intrinsic organization and their extrinsic functional connections, are not assigned particular 'subtasks'. One recalls for instance that important 'logics' for recognition memory seem to reside in the area of the anterior thalamus and that the hypothalamically induced rage can be facilitated or inhibited by lesions or stimulation in the dorsolateral and the ventromedial amygdala, respectively.

Fourth, some of these 'limbic symptoms', i.e. changes in behavior after various limbic manipulations, appear to manifest some degree of 'abnormality' if not 'neuropsychopathology'. This holds true certainly for such phenomena as the ('unfounded') anger, as induced by stimulation of the perifornical area, the undiscriminative sexual behavior after limbic lesions and the unwarranted seeking of 'pleasure' in the self-stimulation experiment where 'all caution and consideration for other needs (and dangers) is completely forgotten.' This holds true, too, for the obvious disturbances in the memory functions as well as for the signs of visual agnosia *and* the oral tendencies.

If we make an attempt towards an interpretation of all the findings discussed here, we may state that the animal's limbic system – probably through a well-balanced operation of its various substructures – is involved in the organization and control of the various moods, drives, and motivational states and the behaviors resulting from these moods and drives. In addition it seems to control – probably, though, not to organize in detail – various memory functions. It seems to produce, again in a delicately balanced way, that particular degree of behavioral inhibition that is necessary for a correct and adequately equilibrated behavioral output. And finally, it is also involved in the organization of sleep, and in particular of that local vigilant state we refer to as REM sleep, where everything (or almost everything) is at rest while those systems that let us live through, and remember, a 'happening' are highly active.

Some, yet not all, of the changes and behavioral abnormalities produced by experimental intervention in the limbic system of animals imitate the symptoms we are used to see in man suffering from schizophrenia, mania or complex partial epileptic seizures. It seems that animals cannot exactly imitate man's symptoms as the latter are defined in terms of the specific human armamentarium of behaviors and experiential possibilities, including verbal communication. But we may start to identify on the basis of the 'limbic symptomatology' in animals some basic functions handled by this part of the brain, postulate that these very same functions are handled, mutatis mutandis, by the human limbic system and then make an attempt to relate man's symptomatology to these functions. If we succeed in doing so for various symptom groups in schizophrenics, manics and (temporal lobe) epileptics, we have good reason to postulate a (in part at least) common pathology of the three afflictions. Experimental medicine and morphological as well as functional neuropathology should be able, and willing, to test this postulate.

References

1. Adamec, R.E.: Hypothalamic and extrahypothalamic substrates of predatory attack. Suppression and the influence of hunger. Brain Res. *106:* 57–69 (1976).
2. Adey, W.R.; Walter, D.O.: Application of phase detection in the cat. Expl. Neurol. *7:* 186–209 (1963).
3. Anand, B.K.; Dua, S.: Electrical stimulation of the limbic system of brain ('visceral brain') in waking animals. Indian J. med. Res. *44:* 107–119 (1956).
4. Arbuthnott, G.W.; Fuxe, K.; Ungerstedt, U.: Central catecholamine turnover and self-stimulation behavior. Brain Res. *27:* 406–413 (1971).
5. Benardo, L.S.; Prince, D.A.: Acetylcholine-induced modulation of hippocampal pyramidal neurons. Brain Res. *211:* 227–234 (1981).
6. Bischoff, S.; Bittiger, H.; Krauss, J.: In vivo [^{3}H] spiperone binding to the rat hippocampal formation: involvement of dopamine receptors. Eur. J. Pharmacol. *68:* 305–315 (1980).
7. Black, A.H.; Nadel, L.; O'Keefe, J.: Hippocampal function in avoidance learning and punishment. Psychol. Bull. *84:* 1107–1129 (1977).
8. Black, A.H.; Young, G.A.: Electrical activity of the hippocampus and cortex in dogs operantly trained to move and to hold still. J. comp. physiol. Psychol. *79:* 128–141 (1972).
9. Brooks, F.P.: Central neural control of acid secretion; in Heidel, Code, Handbook of physiology; Vol. II, pp. 805–826 (American Physiological Society, Washington, D.C. 1967).
10. Bucy, P.C.; Klüver, H.: An anatomical investigation of the temporal lobe in the monkey *(Macaca mulatta).* J. comp. Neurol. *103:* 151–251 (1955).
11. Buijs, R.M.: Immunocytochemical demonstration of vasopressin and oxytocin in the rat brain by light and electron microscopy. J. Histochem. Cytochem. *28:* 357–360 (1980).
12. Crow, T.J.: A map of the rat mesencephalon for electrical self-stimulation. Brain Res. *36:* 265–273 (1972).
13. Crow, T.J.: The neuroanatomy of intracranial self-stimulation: a generel catecholamine hypothesis. Neurosci. Res. Program Bull. *15:* 195–204 (1977).
14. Daruna, J.H.; Kent, E.W.: Comparison of regional serotonin levels and turnover in the brain of naturally high and low aggressive rats. Brain Res. *101:* 489–501 (1976).
15. Deutsch, J.A.: The cholinergic synapse and the site of memory. Science *174:* 788–794 (1971).

16 de Wied, D.: Behavioral actions of neurohypophysial peptides. Proc. R. Soc., London ser. B. *210:* 183–195 (1980).
17 Dreyfuss, J.J.; Mühlethaler, M.: Vasopressin-induced modulation of firing rate in hippocampal slices. J. Physiol., Lond. *317:* 28 (1981).
18 Duka, T.; Wüster, M.; Schubert, P.; Stoiber, R.; Herz, A.: Selective localization of different types of opiate receptors in hippocampus as revealed by in vitro autoradiography. Brain Res. *205:* 181–186 (1981).
19 Egger, M.D.; Flynn, J.P.: Effects of electrical stimulation of the amygdala on hypothalamically elicited attack behavior in cats. J. Neurophysiol. *26:* 705–720 (1963).
20 Epelbaum, J.; Tapia-Arancibia, L.; Kordon, C.: Noradrenaline stimulates somatostatin release from incubated slices of the amygdala and the hypothalamic preoptic area. Brain Res. *215:* 393–397 (2981).
21 Flood, J.F.; Landry, D.W.; Jarvik, M.E.: Cholinergic receptor interactions and their effects on long-term memory processing. Brain Res. *215:* 177–185 (1981).
22 Fonberg, E.; Delgado, J.M.R.: Avoidance and alimentary reactions during amygdala stimulation. J. Neurophysiol. *24:* 651–664 (1961).
23 Fuller, J.L.; Rosvold, H.E.; Pribram, K.H.: The effect on affective and cognitive behavior in the dog of lesions of the pyriform-amygdala-hippocampal complex. J. comp. physiol. Psychol. *50:* 89–96 (1957).
24 German, D.C.; Bowden, D.M.: Catecholamine systems as the neural substrate for intracranial self-stimulation: a hypothesis. Brain Res. *73:* 381–419 (1974).
25 Goddard, G.V.: Functions of the amygdala. Psychol. Bull. *62:* 89–109 (1964).
26 Gotsick, J.E.; Drew, W.G.; Proctor, D.L.: Apomorphine-induced aggression: an evaluation of possible sensitizing factors in the rat. Pharmacology *13:* 385–390 (1975).
27 Grastyan, E.; Karmos, G.: The influence of hippocampal lesions on simple and delayed instrumental conditioned reflexes; in Physiologie de l'hippocampe. Colloques int. Centr. natn. Rech. scient. *107:* 225–239 (1962).
28 Grastyan, E.; Lissàk, K.; Madarász, I.; Donhoffer, H.: Hippocampal electrical activity during the development of conditioned reflexes. Electroenceph. clin. Neurophysiol. *11:* 409–430 (1959).
29 Gray, J.; Feldon, J.; Rawlins, J.N.P.; Owen, S.; McNaughton, N.: The role of the septo-hippocampal system and its noradrenergic afferents in behavioral responses to non-reward; in Elliott, Whelan, Functions of the septo-hippocampal system, CIBA Foundation Symposium 58 (new series), pp. 275–300 (Elsevier, Amsterdam 1978).
30 Green, J.D.; Clemente, C.D.; Groot, J. de: Rhinencephalic lesions and behavior in cats. J. comp. Neurol. *108:* 505–545 (1957).
31 Grijalva, C.V.; Lindholm, E.; Novin, D.: Physiological and morphological change in the gastrointestinal tract induced by hypothalamic intervention: an overview. Brain Res. Bull. *5:* suppl. 1, pp. 19–31 (1980).
32 Grinberg-Zylberbaum, J.; Carranza, M.B.; Cepeda, G.V.; Vale, T.C.; Steinberg, N.N.: Caudate nucleus stimulation impairs the process of perceptual integration. Physiol. Behav. *12:* 913–918 (1974).
33 Hess, W.R.: Das Zwischenhirn (Schwabe, Basel 1949)
34 Hoebel, B.: Hypothalamic self-stimulation and stimulation escape in relation to feeding and drinking. Fed. Proc. *38:* 2454–2461 (1979).
35 Hoebel, B.G.; Teitelbaum, P.: Hypothalamic control of feeding and self-stimulation. Science *135:* 375–377 (1962).

36 Hunsperger, R.W.: Affektreaktionen auf elektrische Reizung im Hirnstamm der Katze. Helv. physiol. Acta *14:* 70–92 (1956).
37 Hunsperger, R.W.: Effects of brain lesions on prey-catching and male sexual behavior in the domestic cat. J. Physiol., Lond. *317:* 1 (1981).
38 Ito, M.; Olds, J.: Unit activity during self-stimulation behavior J. Neurophysiol. *34:* 263–273 (1971).
39 Jacobs, B.L.; McGinty, D.J.: Participation of the amygdala in complex stimulus recognition and behavioral inhibition: evidence from unit studies. Brain Res. *36:* 431–436 (1972).
40 Jonason, K.R.; Enloe, L.J.; Contrucci, J.; Meyer, P.M.: Effects of simultaneous and successive septal and amygdaloid lesions on social behavior of the rat. J. comp. physiol. Psychol. *83:* 54–61 (1973).
41 Jouvet, M.: The rhombencephalic phase of sleep; in Moruzzi, Fessard, Jasper, Progr. Brain Res., vol. 1: Brain mechanisms, pp. 407–424 (Elsevier, Amsterdam 1963).
42 Kaada, B.R.: Somato-motor, autonomic and electrocorticographic responses to electrical stimulation of 'rhinencephalic' and other structures in primates, cat, and dog. Acta physiol. scand. *24:* suppl. 83, pp. 1–285 (1951).
43 Kaada, B.R.; Rasmussen, E.W.; Kveim, O.: Impaired acquisition of passive avoidance behavior by subcallosal, septal, hypothalamic, and insular lesions in rats. J. comp. physiol. Psychol. *55:* 661–670 (1962).
44 Kawakami, M.; Kimura, F.: Limbic-preoptic responses to estrogens and catecholamines in relation to cyclic LH secretion; in Naftolin, Ryan, Davies, Subcellular mechanisms in reproductive neuroendocrinology, pp. 423–452 (Elsevier, Amsterdam 1976).
45 Kawakami, M.; Terasawa, E.; Kimura, F.; Wakabayashi, K.: Modulating effect of limbic structures on gonadotrophin release. Neuroendocrinology *12:* 1–16 (1973).
46 Kelly, P.H.; Seviour, P.W.; Iversen, S.D.: Amphetamine and apomorphine responses in the rat following 6-OHDA lesions of the nucleus accumbens septi and corpus striatum. Brain Res. *94:* 507–522 (1975).
47 Kemp, I.R.; Kaada, B.R.: The relation of hippocampal theta activity to arousal, attentive behaviour and somato-motor movements in unrestrained cats. Brain Res. *95:* 323–342 (1975).
48 Kennard, M.; The cingulate gyrus in relation to consciousness. J. nerv. ment. Dis. *121:* 34–39 (1955).
49 Kimble, D.P.: Possible inhibitory functions of the hippocampus. Neuropsychologia *7:* 235–244 (1969).
50 Kimble, D.P.; Jacobson-Kimble, R.: The effect of hippocampal lesions on extinction and 'hypothesis' behavior in rats. Physiol. Behav *5:* 735–738 (1970).
51 Kimura, H.; Kuriyama, K.: Distribution of gamma amino butyric acid (GABA) in the rat hypothalamus; function correlates of GABA with activities of appetite controlling mechanisms. J. Neurochem. *24:* 903–907 (1975).
52 King, F.A.; Meyer, P.M.: Effects of amygdaloid lesions upon septal hyperemotionality in the rat. Science *128:* 655–656 (1958).
53 Kling, A.; Orbach, J.; Schwartz, N.B.; Towne, J.C.: Injury to the limbic system and associated structures in cats. Archs gen. Psychiat. *3:* 391–420 (1960).
54 Kling, A.; Schwartz, N.B.: Effects of amygdalectomy on sexual behaviour and reproductive capacity in the male rat. Fed. Proc. *20:* 335 (1961).

55 Kling, A.; Steklis, H.D.; Deutsch, S.: Radiotelemetered activity from the amygdala during social interactions in the monkey. Expl. Neurology 66: 88–96 (1979).
56 Klüver, H.: Brain mechanisms and behavior with special reference to the rhinencephalon. Lancet 72: 567–574 (1952).
57 Klüver, H.; Bucy, P.C.: 'Psychic blindness' and other symptoms following bilateral temporal lobectomy in Rhesus monkeys. Am. J. Physiol. 119: 352–353 (1937).
58 Klüver, H.; Bucy, P.C.: Preliminary analysis of functions of the temporal lobe in monkeys. Archs Neurol. Psychiat. 42: 979–1000 (1939).
59 Koella, W.P.: Vigilance – a concept and its neurophysiological and biochemical implications; in Passouant, Oswald, Pharmacology of the states of alertness, (pp. 171–178 (Pergamon Press, New York 1979).
60 Koella, W.P.: Neurotransmitters and sleep; in Wheatley, Psychopharmacology of sleep, pp. 19–52 (Raven Press, New York 1981).
61 Koella, W.P.; Feldstein, A.; Czicman, J.: The effect of para-chlorophenyl-alanine on the sleep of cats. Electroenceph. clin. Neurophysiol. 25: 481–490 (1968).
62 Krieckhaus, E.E.; Simmons, H.J.; Thomas, G.J.; Kenyon, J.: Septal lesions enhance shock avoidance behavior in the rat. Expl Neurol. 9: 107–113 (1964).
63 Landfield, P.W.: Different effects of posttrial driving or blocking of the theta rhythm on avoidance learning in rats. Physiol. Behav. 18: 439–445 (1977).
64 Lewińska, M.K.: Changes in eating and drinking produced by partial amygdalar lesions in cat. Bull. Acad. pol. Sci. 15: 301–305 (1967).
65 Lewis, P.R.; Shute, C.C.D: The cholinergic limbic system: projections to hippocampal formation, medial cortex, nuclei of the ascending cholinergic reticular system, and the subfornical organ and supra-optic crest. Brain 90: 521–540 (1967).
66 Lidsky, T.I.; Levine, M.S.; MacGregor, Jr.: Hippocampal units during orienting and arousal in rabbits. Expl Neurol. 44: 171–186 (1974).
67 Lopes da Silva, F.H., Kamp, A.: Hippocampal theta frequency shifts and operant behavior. Electroenceph. clin. Neurophysiol. 26: 133–143 (1969).
68 Lorens, S.A.; Brown, T.S.: Influence of stimulation of the septal area on visual evoked potentials. Expl. Neurol. 17: 86–90 (1967).
69 MacLean, P.D.: The limbic system and its hippocampal formation. Studies in animals and their possible application to man. J. Neurosurgery 11: 29–44 (1954).
70 MacLean, P.D.: The limbic system ('visceral brain') and emotional behavior. Archs Neurol. Psychiat. 73: 130–134 (1955).
71 MacLean, P.D.: Contrasting functions of limbic and neocortical systems of the brain and their relevance to psychophysiological aspects of medicine. Am. J. Med. 25: 611–626 (1958).
72 MacLean, P.D.; Ploog, D.W.: Cerebral representation of penile erection. J. Neurophysiol. 25: 29–55 (1962).
73 Margules, D.L.: Alpha and beta adrenergic receptors in amygdala: reciprocal inhibitors and facilitators of punished operant behavior. Eur. J. Pharmacol. 16: 21–26 (1971).
74 Mark, V.H.; Takada, I.; Tsutsumi, H.; Takamatsu, H.; Toth, E.; Mark, D.B.: Effect of exogenous catecholamines in the amygdala of a 'rage' cat. Appl. Neurophysiol. 38: 61–72 (1975).
75 McKenzie, G.M.: Role of tuberculum olfactorium in stereotyped behaviour induced by apomorphine in the rat. Psychopharmacologia 23: 212–219 (1972).

76 McLain, W.C.; Cole, B.T.: Central catechol- and indolamine systems and aggression. Pharmacol. Biochem. Behav. *2:* 123–126 (1974).
77 Means, L.W.; Leander, J.D.; Isaacson, R.L.: The effects of hippocampectomy on alternation behavior and response to novelty. Physiol. Behav. *6:* 17–22 (1971).
78 Mirsky, A.F.; Rosvold, H.E.; Pribram, K.H.: Effects of cingulectomy on social behavior in monkeys. J. Neurophysiol. *20:* 588–601 (1957).
79 Mondadori, C.; Waser, P.G.: Facilitation of memory processing by posttrial morphine: possible involvement of reinforcement mechanisms. Psychopharmacology *63:* 297–300 (1979).
80 Moore, R.Y.; Bloom, F.E.: Central catecholamine neuron systems: anatomy and physiology of the dopamine systems. Annu. Rev. Neurosci. *1:* 129–169 (1978).
81 Moore, R.Y.; Bloom, F.E.: Central catecholamine neuron systems: anatomy and physiology of the norepinephrine and epinephrine systems. Annu. Rev. Neurosci. *2:* 113–168 (1979).
82 Morales, F.R.; Roig, J.A.; Monti, J.M.; Macadar, O.; Budelli, R.: Septal unit activity and hippocampal EEG during the sleep-wakefulness cycle in the rat. Physiol. Behav. *6:* 563–567 (1971).
83 Morley, J.E.; Levine, A.S.; Silvis, S.E.: Endogenous opiates inhibit gastric acid secretion induced by central administration of thyrotropin-releasing hormone (TRH). Life Sci. *29:* 293–297 (1981).
84 Nakao, H.: Hypothalamic emotional reactivity after amygdaloid lesions in cats. Folia psychiat. neurol. jap. *14:* 357–366 (1960).
85 Nauta, W.J.H.: Hippocampal projections and related neural pathways to the midbrain in the cat. Brain *81:* 319–340 (1985).
86 Niki, H.: Effects of hippocampal ablation on learning in the rat; in Adey, Tokizane, Progr. Brain Res. vol. 27: Structure and function of the limbic system, pp. 305–317 (Elsevier, Amsterdam 1967).
87 Nishizuka, M.; Arai, Y.: Sexual dimorphism in synaptic organization in the amygdala and its dependence on neonatal hormone environment. Brain Res. *212:* 31–38 (1981).
88 O'Keefe, J.; Black, A.H.: Single unit and lesion experiments on the sensory inputs to the hippocampal cognitive map; in Elliott, Whelan, Functions of the septo-hippocampal system. Ciba Foundation Symp. 58 (new series) pp. 169–192 (Elsevier, Amsterdam 1978).
89 O'Keefe, J.; Dostrovsky, J.: The hippocampus as a spatial map. Preliminary evidence from unit activity in the freely-moving cat. Brain Res. *34:* 171–175 (1971).
90 Olds, J.; Hirano T.: Conditioned responses of hippocampal and other neurons. Electroenceph. clin. Neurophysiol. *26:* 159–166 (1969).
91 Olds, J.; Milner, P.M.: Positive reinforcement produced by electrical stimulation of the septal area and other regions of the rat brain. J. comp. physiol. Psychol. *47:* 419–427 (1954).
92 Olton, D.S.: The functions of septo-hippocampal connections in spatially organized behavior; in Elliott, Whelan, Functions of the septo-hippocampal system. Ciba Foundation Symp. 58 (new series), pp. 327–342 (Elsevier, Amsterdam 1978).
93 Olton, D.S.; Issacson, R.L.: Fear, hippocampal lesions and avoidance behavior. Commun. behav. Biol. *3:* 259–262 (1969).
94 Olton, D.S.; Walker, J.A.; Gage, F.H.: Hippocampal connections and spatial discrimination. Brain Res. *139:* 295–308 (1978).

95 Orbach, J.; Milner, B.; Rasmussen, T.: Learning and retention in monkeys after amygdala-hippocampus resection. Archs Neurol. *3:* 230–251 (1960).
96 Papez, J.W.: A proposed mechanism of emotion. Archs Neurol. Psychiat. *38:* 725–743 (1937).
97 Parmeggiani, P.L.: Schlafverhalten bei repetierender elektrischer Reizung im Thalamus, Mittelhirn und Fornix an der nichtnarkotisierten Katze. Pflügers Arch. ges. Physiol. *174:* 84–85 (1961).
98 Peroutka, S.J.; Snyder, S.H.: Two distinct serotonin receptors: regional variations in receptor binding in mammalian brain. Brain Res. *208:* 339–347 (1981).
99 Poschel. B.P.H.: The hippocampally stimulated rat as a model for Korsakoff-type amnesias. Lab. Anim. Sci. *27:* 738–747 (1977).
100 Raisman, C.; Field, P.M.: Sexual dimorphism in the preoptic area of the rat. Science *173:* 731–733 (1971).
101 Ray, A.; Sharma, K.K.; Sen, P.: Effects of histaminergic drugs on footshock-induced aggressive behaviour in rats. Eur. J. Pharmacol. *73:* 217–219 (1981).
102 Rolls, T.: Processing beyond the inferior temporal visual cortex related to feeding, memory, and striatal function; in Katsuki, Norgren, Sato, Brain mechanisms of sensation, pp. 241–269 (Wiley & Sons, 1981).
103 Rosvold, H.E.; Mirsky, A.F.; Pribram, K.H.: Influence of amygdalectomy on social behavior in monkeys. J. comp. physiol. Psychol. *47:* 173–178 (1954).
104 Routtenberg, A.: Intracranial self-stimulation: catecholamine brain pathways and memory consolidation; in Cole, Sonderegger, Nebraska Symp. on Motivation, pp. 161–182 (University of Nebraska Press, Lincoln, Nebr. 1975).
105 Routtenberg, A.: Participation of brain stimulation reward substrates in memory: anatomical and biochemical evidence. Fed. Proc. *38:* 2446–2453 (1979).
106 Routtenberg, A.; Malsbury, C.: Brainstem pathways of reward. J. comp. physiol. Psychol. *68:* 22–30 (1969).
107 Rozé, C.; Dubrasquet, M.; Chariot, J.; Vaille, C.: Central inhibition of basal pancreatic and gastric secretions by β-endorphin in rats. Gastroenterology *79:* 659–664 (1980).
108 Sawyer, C.H.: Triggering of the pituitary by the central nervous system; in Bullock, Physiological triggers, pp. 164–174 (Waverly Press, Baltimore 1957).
109 Schreiner, L.; Kling, A.: Behavioral changes following rhinencephalic injury in the cat. J. Neurophysiol. *16:* 643–659 (1953).
110 Schreiner, L.; Kling, A.: Rhinencephalon and behavior. Am. J. Physiol. *184:* 486–490 (1956).
111 Siegel, A.; Edinger, H.; Dotto, M.: Effects of electrical stimulation of the lateral aspects of the prefrontal cortex upon attack behavior in cats. Brain *93:* 473–484 (1975).
112 Stein, L.; Belluzzi, J.D.: Brain endorphins: possible role in reward and memory formation. Fed. Proc. *38:* 2468–2472 (1979).
113 Stein, L.; Wise, C.D.: Release of norepinephrine from hypothalamus and amygdala by rewarding medial forebrain bundle stimulation and amphetamine. J. comp. physiol. Psychol. *67:* 189–198 (1969).
114 Stengaard-Pedersen, K.; Fredens, K.; Larsson, L.-I.: Enkephalin and zinc in the hippocampal mossy fiber system. Brain Res. *212:* 230–233 (1981)
115 Sternberg, D.B.; Gold, P.E.: Retrograde amnesia produced by electrical stimulation of the amygdala: attenuation with adrenergic antagonists. Brain Res. *211:* 59–65 (1981).

116 Stevens, J.R.: Psychomotor epilepsy and schizophrenia: a common anatomy? in Brazier, Epilepsy, its phenomena in man, pp. 189–214 (Academie Press, New York 1973).
117 Taché, Y.; Vale, W.; Rivier, J.; Brown, M.: Brain regulation of gastric secretion: influence of neuropeptides. Proc. natn. Acad. Sci. 77: 5515–5519 (1980).
118 Ursin, H.; Kaada, B.R.: Functional localization within the amygdaloid complex in the cat. Electroenceph. clin. Neurophysiol. 12: 1–20 (1960).
119 Vergnes, M.; Karli, P.: Effet de l'ablation des bulbes olfactifs et de l'isolement sur le développement de l'agressivité interspécifique du rat. Soc. Biol. 163: 2704 (1969).
120 Vincent, S.R.; McGeer, E.G.: A substance P projection to the hippocampus. Brain Res. 215: 349–351 (1981).
121 Weiskrantz, L.; Warrington, E.K.: The problem of the amnestic syndrome in man and in animals; in Isaacson, Pribram, The hippocampus, vol. 2: Neurophysiology and behavior pp. 411–428 (Plenum Press, New York 1975).
122 Wise, R.A.: Catecholamine theories of reward: a critical review. Brain 152: 215–247 (1978).
123 Wise, R.A.: Intracranial self-stimulation: mapping against the lateral bounderies of the dopaminergic cells of the substantia nigra. Brain Res. 213: 190–194 (1981).
124 Wolf, P.; Haas, H.L.: Effects of diazepines and barbiturates on hippocampal recurrent inhibition. Arch. Pharmacol. 299: 211–218 (1977).
125 Wood, C.D.: Behavioral changes following discrete lesions of temporal lobe structures. Neurology 8: 215–220 (1958).
126 Zieglgänsberger, W.; French, E.D.; Siggins, G.R.; Bloom, F.E.: Opioid peptides may excite hippocampal pyramidal neurons by inhibiting adjacent inhibitory interneurons. Science 205: 415–417 (1979).

W.P. Koella, MD, Friedrich Miescher Institute, CH-4002 Basel (Switzerland)

Synaptic Control and Modulation of Hippocampal Neurones

Per Andersen

Institute of Neurophysiology, University of Oslo, Oslo, Norway

Introduction

The hippocampus is part of the limbic system [50]. It represents a most regularly organized cortical region.

In recent years, great interest has focused upon the hippocampus for several reasons. First, interference with the hippocampus or structures near it has given rise to memory deficits in man [50]. Second, animal experiments have indicated that the hippocampus may be important for spatial memory [68] or tasks requiring nonspatial working memory [57, 58]. Third, the hippocampus has been proposed to be involved in certain mental states [50] including disorders associated with depression. The basis for this idea has been its connection with the amygdala and certain regions and its neuromodulator content. Finally, a major reason for much research has been its great degree of order which makes it a good model for neurobiological studies.

The overall position of the hippocampus in the rodent brain is indicated in figure 1A. The outline of the hippocampus is drawn superimposed on the lateral aspect of the brain. In rodents, the hippocampus comprises a relatively large part of the volume of the brain (10–13%). On ascending the phylogenetic scale the relative volume decreases while the absolute volume increases, being highest in man and whales. Dorsally and anteriorly it borders the septal region whereas the amygdaloid nuclear complex is found anterior and medial to its temporal pole. Caudally, the hippocampus borders the entorhinal area which is a part of the hippocampal gyrus.

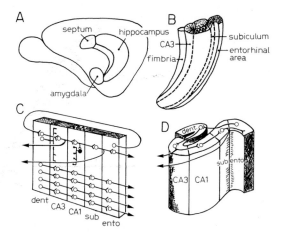

Fig. 1. Diagram of the hippocampal organization. *A* The outline of the left hippocampus superimposed upon the left hemisphere. *B* The temporal part of the hippocampal formation with subdivisions seen in the same projection as in *A*. *C* Simplified and unfolded hippocampal formation with subdivisions and main neuronal arrangement. The two curved arrows are outputs to the septum and the hypothalamus. The arrows on the right are outputs to neocortical regions. *D* The same elements as in *C*, but now folded as in the brain. Same angle of view as in *B*. dent = dentate area; sub = subiculum, ento = entorhinal area.

The Organization of Hippocampal Formation

Inside the hippocampus, one is struck by the high degree of order and the uniformity throughout the structure. Figure 1B shows how the hippocampal cortex is curved twice, first around a line from the septum to the temporal tip (septotemporal axis) and, second, around a line normal to this axis, to a C-shaped form. The different cortical elements which border each other to compose the hippocampal formation are indicated with different symbols. Comparison of these symbols with the unfolded cortical sheet, shown in figure 1C, should indicate how the cortex is composed of at least five parallel strips bordering each other [47]. These strips then fold around a line parallel to their long sides (fig. 1B, D). Figure 1C and D are meant to indicate how the input fibers enter the entorhinal region. The next set of fibers passes across the obliterated hippocampal fissure. The main pathway for impulses in the hippocampal formation is relayed by four serial elements (fig. 1D). The fiber systems are in sequence: perforant path fibers; mossy fibers; CA3 Schaffer collaterals; CA1 axons, and subicular axons. The output fibers are collateral branches of the CA3 neurones which course to the fimbria and fibers from the subiculum, which run in the dorsal fornix. These fibers reach the septal region and various basal forebrain areas, notably different parts of the hypothalamus [61]. Near the temporal end, certain output fibers reach the amygdaloid nuclear complex. In all parts of the hippocampal formation, the fibers comprising the main impulse system remain parallel to each other to give the system an orderly appearance (fig. 1D). The axons have, however, many collaterals which may excite other neurones. One neurone is the polymorph neurone which sends axons longitudinally [28]. Another example is interneurones, one of which is drawn in figure 1C as a black structure. Because of this arrangement, a thin slice taken at right angles to the septotemporal axis, will contain all major elements and thus comprise a micro-hippocampus or a lamella [4, 12].

The organization of the hippocampus is characterized by the principles of lamellar arrangement, stratification, orthogonal input and morphologic distinction between synapses of different function.

Stratification of Synapses

A major organizational rule is that the afferent fibers course parallel to each other and to the main cellular layer. In the hippocampus, the nerve cells are gathered in a single layer (stratum pyramidale) with their apical and basal dendrites found in stratum radiatum and oriens, respectively [47, 64]. This means that afferent fibers from a given source end on a specific part of the dendritic tree with fibers crossing the dendritic tree orthogonally. At the crossing, excitatory fibers make *en passage* boutons nearly exclusively with spines [3]. Contacts with interneurones are made directly upon their dendritic shafts which often have a beaded appearance. In contrast, interneurones, of which there are several in the hippocampal cortex, tend to make synapses on the soma and on the largest dendrites and their closest branches. The excitatory synapses are of the asymmetric type, whereas the interneurones seem to terminate with mostly symmetric synapses. A particularly interesting synapse is the coupling between the dentate granule cells and the CA3 pyramids. Here, the thin mossy fibers enlarge to large globules into which finger-like processes of the initial part of the CA3 apical dendrites burrow. In this synapse there are multiple active sites in the one giant bouton. The vesicles are round and very numerous. In addition, large mitochondriae and a few dense core vesicles are usually found. On the whole, excitatory synapses usually have round vesicles, whereas inhibitory boutons often may have pleomorphic vesicles.

Synaptic Aspects

The arrival of the isolated cortical slices [74, 86] represents a great advantage. Many neurobiological questions can now be answered with a direct approach. Thus, we now know that the major effect of hippocampal synapses are chemically mediated transmission. The identity of the excitatory transmitters are unknown as yet. However, a good candidate in nearly all portions of the hippocampal cortex is glutamate [10, 24, 53, 54, 67, 75, 77, 78, 81]. In certain systems, aspartate may be the transmitter [45, 53]. Judged from the effect of iontophoretically applied glutamate [10, 24, 36, 67], glutamate-sensitive areas are found in all major portions of the dendritic tree, with the possible exception of the terminal fifth.

Synaptic Power

Recalling the large number of excitatory synapses on one cell, one would assume that the most distal synapses would have considerably less effect than more proximal synapses [62, 63]. However, when afferent fibers located at various distances from the soma were excited, the EPSPs produced were remarkably similar in time course (fig. 2). Furthermore, when the input synaptic current was controlled for equal magnitude, the probability of discharge was very similar for proximal and distal synaptic inputs [7]. Furthermore, when lesions were made to abolish all synaptic inputs, except a small bundle of fibers crossing a tissue bridge at a known location, a very small portion of afferent fibers was sufficient to drive the cell provided they were activated synchronously. For CA1 cells the required number of fibers to drive the cell with the probability of 1.0 was estimated to 3% of the total amount of afferent fibers to the cell. Furthermore, when such lesions were made at different levels from the soma, all fiber bundles of the same magnitude were equally efficient [7].

Such a remarkable presynaptic efficiency could be explained by the amount of transmitter released at each bouton being large or by an effective summation of neighboring activated synapses, or a mixture of the two. Single fiber EPSPs have only been measured in the perforant path/granule cell synapses. Here three different methods gave values around 0.1 mV [51]. The degree of nonlinearity of the EPSP summation was estimated from the size of the extracellular EPSP related to the input strength. With

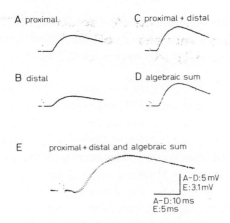

Fig. 2. A Average EPSP of CA1 pyramidal cell (n = 30) to selective activation of fibers synapsing on proximal parts of the apical dendrite. *B* Corresponding EPSP to a distal dendritic input. *C* EPSP due to the simultaneous activation of the two inputs. *D* Algebraic summation of the EPSPs shown in *A* and *B*. *E* Same responses as in *C* and *D*, but shown superimposed and expanded. EPSPs summate linearly. From *Langmoen and Andersen* [43].

this method, only about 400 synapses needed to be activated simultaneously in order to drive a granule cell. If a similar figure is valid for CA1 pyramidal cells, one would expect an even smaller number of synapses to be necessary since their spike threshold is considerably lower. Because the CA1 pyramidal EPSPs add linearly, a remarkably small figure, only 60 activated synapses, would be necessary to drive the cell. A preliminary investigation has shown that single-fiber EPSP of about 0.1 mV may be recorded from CA1 pyramids.

Synaptic Summation

Since the cells in the hippocampal formation have a large number of afferent synapses, they could conceivably summate nonlinearly by reducing the driving potential for subsequently activated synaptic currents. However, direct observation shows that EPSPs on CA1 pyramidal cells summate linearly (fig. 2C, D) [44]. Figure 2 gives the averaged EPSPs to 30 trials to two separate inputs as described above. The lower trace shows the superimposed responses when both inputs are activated simultaneouly (observed sum) and the algebraic summation of the two individual EPSP (algebraic sum). In about 20% of the cells, these sums were perfectly similar. In the remaining cells, however, there was a difference in that the observed sum was less than the algebraic sum. This deficit turned out to be due to an IPSP, probably of the recurrent variety. Thus, when the experiment was repeated close to the equilibrium potential for the IPSP, the observed sum was similar to the algebraic sum. The conclusion of these experiments is that synaptic potentials are summated linearly whether they are of an excitatory or inhibitory nature. In conclusion, it is not necessary to envisage a very large number of afferent fibers activated synchronously in order to drive the cell, but rather that a small portion of synchronously activated fibers would do this. A pyramidal cell may act as an output device for several afferent sources, provided the source can activate the synapses to the cell in question in a synchronous fashion.

Modification through Interneurones

The hippocampal formation contains a great number of interneurones [47, 64]. So far, only inhibitory interneurones have been found with certainty. The possibility that excitatory interneurones are present remains plausible, however [20]. One type of the inhibitory interneurone, the basket cell, seems to operate in a recurrent loop [6]. Judging from the field potential distribution, the author noted that the synapses were located to the soma of pyramidal cells. Later, *Curtis* et al. [18] maintained that gamma-aminobutyric acid (GABA) was the inhibitory transmitter for the recurrent inhibition. ^{3}H-GABA is taken up in areas where inhibitory synapses are located, and glutamate decarboxylase (GAD), the enzyme producing GABA, is found both around the cell bodies and in the dendritic tree [37, 79]. The GAD content does not change following lesions of major afferent fiber systems and is, therefore, likely to be found in interneurones [76]. Recently, we have recorded intracellularly while GABA was being applied near the soma of hippocampal pyramids. We observed a hyperpolarization associated with a conductance increase [5] (fig. 3). The inhibition is mediated by chloride ions [2]. The equilibrium potential was 9–12 mV negative to the resting potential [22]. *Ribak* et al. [65] have shown by immuno-fluorescence that GAD is found in cells which are situated and terminate exactly as the basket cells.

In addition to the GABA-mediated recurrent inhibition a new inhibitory mechanism was recently found. This new type is also operated by GABA, but with a different type of action. In this case, GABA had to be applied in the dendritic region at specific places. Here, it produces a conductance increase, but associated with a depolarization. The depolar-

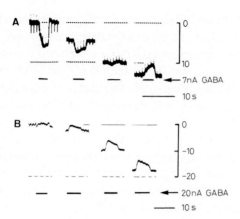

Fig. 3. A Responses of a CA1 pyramidal cell to GABA ejected ionophoretically (7 nA) near its soma. The different sections were obtained at various membrane potentials due to passage of current across the soma membrane. The normally hyperpolarizing response (left) reversed about 10 mV negative to the resting membrane potential. *B* Responses from another CA1 pyramids to GABA (20 nA) ejected among the apical dendrites. The response increased with increasing membrane potential.

ization may, in fact, be so large that it leads to cellular discharge. This effect is also sensitive to the chloride concentration. The equilibrium potential is not known with certainty. However, by extrapolation it appears to be near −40 mV, making it unlikely that chloride is the only ion involved. The major effect of this type of inhibition is to produce a localized conductance increase in the dendritic tree. This will effectively shunt excitatory synaptic currents in the vicinity. Remote synapses, however, will not feel this shunt and may be facilitated by the depolarization produced by this GABA mechanism. The mechanism which inhibits local excitatory synapses and facilitates the more remote ones has been called discriminative inhibition and is proposed as a new mechanism for neuronal control. It remains to be seen whether it operates under physiological circumstances. However, *Alger and Nicoll* [1] have seen in barbiturate-treated slices a long-lasting potential which has many characteristics similar to those produced by GABA itself. The idea is that the afferent impulse has excited interneurones which are presumed to be GABA-containing and which may mediate an effect like the one produced by iontophoretic application. Candidates for such GABA-containing interneurones terminating among the dendrites exist [47, 64, 65].

Cholinergic Influences

From the nucleus of the diagonal band and the medial septal nucleus, fibers run and distribute themselves to large portions of the dentate fascia and the hippocampal cortex. The fibers contain acetylcholinesterase and probably release acetylcholine at their terminals. Stimulation of the system produces an increase in the amplitude and frequency of the hippocampal EEG – the so-called theta activity [17, 40, 59]. In a silent hippocampal cortex, it can induce theta activity de novo. Similarly, eserine augments this activity, whereas atropine may block at least one variety of this electrocorticographic rhythm [82]. Commonly, cholinergic theta activity is seen simultaneous with neocortical desynchronization. Iontophoretic application of acetylcholine to CA1 pyramidal cells gives a slow depolarization and a resistance increase [23]. The most likely mechanism of action is a reduction of the resting potassium conductance, similar to that proposed for the neocortical depolarizing effect of acetylcholine [41].

Modulation by Noradrenaline

The hippocampal cortex receives an important noradrenergic input [13, 80]. Both locus coeruleus stimulation and application of noradrenaline produce cessation of spontaneous and glutamate-driven discharges [71, 72]. By applying noradrenaline iontophoretically to hippocampal pyramidal cells, two mechanisms were seen which may explain the inhibitory effect [44, 70]. There was a moderate hyperpolarization (mean value 2.7 ± 0.4 mV) associated with a conductance increase (fig. 4A), probably for chloride ions [39]. The reduction has relatively slow kinetics and noradrenaline only reduced the slow phases of the depolarizing response, not the initial fast part (fig. 4B). Thus, noradrenaline seems to be able to block slow excitatory events, while the fast transients escape. Such slow calcium-mediated responses may be important for epileptiform activity and for dendritic integration. Experiments with calcium blockers indicate that noradrenaline has its main effect by blocking calcium channels [69].

A similar effect, although less dramatic, is seen by the addition of 5-hydroxytryptamine (serotonin). Here, there is also a hyperpolarization with a moderate conductance increase. Experiments with 4-aminopyridine (4-AP) and K^+-superfusion suggest that serotonin activates a potassium channel [69].

No proper study of a dopamine effect has been made on hippocampal cells. Dopamine fibers are present, however, and have been related to schizophrenia [29, 38].

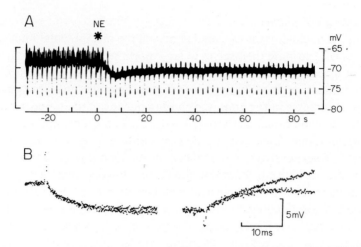

Fig. 4. A Application of a micro-drop (1 nl) of noradrenaline (*, 10^{-3} *M*) gives a small hyperpolarization with a conductance increase. Vertical lines are resistance pulses. *B.* Superimposed responses before and after noradrenaline application to another CA1 pyramidal cell. Responses to depolarizing (right) and hyperpolarizing (left) pulses. The smaller of the two responses to the right is taken under noradrenaline. From *Langmoen* et al. [44].

The hippocampus also contains opioid receptors [33, 35] and enkephalin-containing fibers [31, 66]. The delta receptor is concentrated in the pyramidal layer, predominantly in CA3 close to the CA1 border, while the mu receptor is more widely distributed, including the dendritic areas of CA1 [26]. Enkephalin iontophoresis increases the firing rate of hippocampal pyramids [56]. The mechanism is under dispute. Based on extracellular potentials, *Zieglgänsberger* et al. [87] proposed that reduced inhibition is the cause. This was later supported by others [27, 30, 46, 55]. However, three careful intracellular studies have failed to see any definite changes in inhibitory processes, but have reported increased EPSPs as the main result [21, 34, 48]. Clearly, this fascinating aspect needs further clarification.

Plasticity

Because of its possible relation to memory functions, the observation that hippocampal synapses change their efficiency over time and with different training regimes is interesting. First, dentate granule cells may change their firing frequency in a conditioning paradigm. In freely moving rats, *Deadwyler* et al. [19] were able to find dentate cells which increased their discharge rate subsequent to an auditory signal. Following pairing of the auditory signal with a reward, the latency of discharge and the duration increased. The response could even be redirected when a tone of different frequency was rewarded. After a certain time, the dentate granule cell responded again, but now to the new tone being rewarded. Although the mechanism underlying this phenomenon is not clear, it shows the considerable plasticity of the involved synapses.

More commonly observed is the phenomenon of habituation. In CA3 and CA1, the response to a novel stimulus is followed by an increased discharge [83]. Following continued presentation of the stimulus to the animal with no consequence, there was a decline of the response until no response was detectable. This response could be dishabituated by a strong shock to the paw or a strong sound.

Rearrangement of Afferent Fiber Systems

Several hippocampal fiber systems have the ability to send out collaterals after a neighboring area has been denervated [49, 60]. Particularly effective are the cholinergic and monoaminergic fibers [11, 42], but other fibers also show sprouting. To what extent the new synapses are functional is still not completely certain – although their morphology looks adequate.

An interesting model for plastic changes in the nervous system has been the long-term potentiation first described by *Bliss and Lømo* [15]. A relatively short tetanus within the physiological range (usually 10–20 Hz) gives a long-lasting (hours, even days) potentiation of synaptic transmission in the perforant path/granule cell synapse. Later, similar data have been found for all synapses in the hippocampal formation, where one has looked properly. So far, only excitatory connections have been studied. The phenomenon appears to be homosynaptic in that the majority of investigators agree that only the tetanized pathways show subsequently inreased synaptic efficiency [8, 84]. There are, however, reports to the opposite effect [52]. Trying to find out the mechanisms underlying the effect, studies have been made with intracellular recording from CA1 pyramids [8, 9, 84, 85]. The passive membrane properties of the cells were not changed at all. Furthermore, the input fiber volley was not changed. However, there was a moderate but clear increase of the EPSP, probably indicating an increased amount of transmitter. Indeed, such increased transmitter release has been measured directly [73]. There was also a reduction in the latency of spike discharge and a greatly increased probability of this discharge. These two parameters were more profoundly changed than could be explained by the moderate increase in the EPSP. For this reason it is possible that there are local postsynaptic changes in addition to the presynaptic change with increased transmitter release. This possibility matches the increased phosphorylation of proteins and the occurrence of a specific protein following similar tetanization [16, 25]. Even better potentiations can be seen after repeated short tetani of high frequency (100 Hz for 4 s, repeated once a minute for 3 or 4 times) [32]. Recently, evidence has been presented that noradrenaline must be present for the phenomenon to appear [14].

Although the changes are relatively moderate at a single synapse, the effect in a series of synapses may be considerable. Whether or not the long-term potentiation is an essential element, or only an adjuvant, to the processes underlying learning and memory remains to be seen.

Conclusion

The hippocampal system appears as a useful model for many types of neurobiological studies. Further progress is to be expected and should give us better insight in processes underlying mental phenomena – including mental disorders.

References

1. Alger, B.E.; Nicoll, R.A.: GABA-mediated biphasic inhibitory response in hippocampus. Nature, Lond. *281:* 315–317 (1979).
2. Allen, G.I.; Eccles, J.; Nicoll, R.A.; Oshima, T.; Rubia, F.J.: The ionic mechanisms concerned in generating the i.p.s.ps of hippocampal pyramidal cells. Proc. R. Soc. Ser. B. *198:* 363–384 (1977).
3. Andersen, P.; Blackstad, T.W.; Lømo, T.: Location and identification of excitatory synapses on hippocampal pyramidal cells. Exp. Brain Res. *1:* 236–248 (1966).
4. Andersen, P.; Bliss, T.V.P.; Skrede, K.K.: Lamellar organization of hippocampal excitatory pathways. Exp. Brain Res. *13:* 222–238 (1971).
5. Andersen, P.; Dingledine, R.; Gjerstad, L.; Langmoen, I.A.; Mosfeldt Laursen, A.: Two different responses of hippocampal pyramidal cells to application of gamma-amino butyric acid. J. Physiol., Lond. *305:* 279–296 (1980).
6. Andersen, P.; Eccles, J.C.; Løyning, Y.: Pathway of postsynaptic inhibition in the hippocampus. J. Neurophysiol. *27:* 608–619 (1964).
7. Andersen, P.; Silfvenius, H.; Sundberg, S.H.; Sveen, O.: A comparison of distal and proximal dendritic synapses on CA1 pyramids in hippocampal slices in vitro. J. Physiol., Lond. *307:* 273–299 (1980).

8   Andersen, P.; Sundberg, S.H.; Sveen, O.; Wigström, H.: Specific long-lasting potentiation of synaptic transmission in hippocampal slices. Nature, Lond. *266:* 736–737 (1977).
9   Andersen, P.; Sundberg, S.H.; Swann, J.N.; Wigström, H.: Possible mechanisms for long-lasting potentiation of synaptic transmission in hippocampal slices from guinea pigs. J. Physiol., Lond. *302:* 463–482 (1980).
10 Biscoe, T.J.; Straughan, D.W.: Micro-electrophoretic studies of neurones in the cat hippocampus. J. Physiol., Lond. *183:* 341–359 (1966).
11 Bjørklund, A.; Stenevi, U.; Svendgaard, N.-A.: Growth of transplanted monoaminergic neurones into the adult hippocampus along the perforant path. Nature, Lond. *262:* 787–790 (1976).
12 Blackstad, T.W.; Brink, K.; Hem, J.; Jeune, B.: Distribution of hippocampal mossy fibers in the rat. An experimental study with silver impregnation methods. J. comp. Neurol. 38: 433–450 (1970).
13 Blackstad, T.W.; Fuxe, K.; Hökfelt, T.: Noradrenaline nerve terminals in the hippocampal region of the rat and the guinea pig. Z. Zellforsch. mikrosk. Anat. *78:* 463–473 (1967).
14 Bliss, T.V.P.; Goddard, G.V.; Robertson, H.A.; Sutherland, R.J.: Noradrenaline depletion reduces long term potentiation in the rat hippocampus; in Cellular analogues of conditioning and neural plasticity. Advances in physiological sciences, vol. 36 (Pergamon Press, Oxford 1981).
15 Bliss, T.V.P.; Lømo, T.: Long-lasting potentiation of synaptic transmission in the dentate area of the anaesthetized rabbit following stimulation of the perforant path. J. Physiol., Lond. *232:* 331–356 (1973).
16 Browning, M.; Dunwiddie, T.; Bennett, W.; Gispen, W.; Lynch, G.: Synaptic phosphoproteins: specific changes after repetitive stimulation of the hippocampal slice. Science *203:* 60-62 (1979).
17 Brücke, F.; Petsche, H.; Pillat, B.; Deisenhammer, E.: Ein Schrittmacher in der medialen Septumregion des Kaninchengehirnes. Pflügers Arch. ges. Physiol. *269:* 135–140 (1959).
18 Curtis, D.R.; Felix, D.; McLennan, H.: GABA and hippocampal inhibition. Br. J. Pharmacol. *40:* 881–883 (1970).
19 Deadwyler, S.A.; West, M.; Lynch, G.: Activity of dentate granule cells during learning: differentiation of perforant path input. Brain Res. *169:* 29–43 (1979).
20 Dichter, M.; Spencer, W.A.: Penicillin-induced interictal discharges from the cat hippocampus. I. Characteristics and topographical features. J. Neurophysiol. *32:* 649–662 (1969).
21 Dingledine, R.: Possible mechanisms of enkephalin action on hippocampal CA1 pyramidal neurons. J. Neurosci. *1:* 1022–1035 (1981).
22 Dingledine, R.; Langmoen, I.A.: Conductance changes and inhibitory actions of hippocampal recurrent IPSPs. Brain Res., *185:* 277–287 (1980).
23 Dodd, J.; Dingledine, R.; Kelly, J.S.: The excitatory action of acetylcholine on hippocampal neurones of the guinea pig and rat maintained in vitro. Brain Res. *207:* 109–127 (1981).
24 Dudar, J.D.: In vitro excitation of hippocampal pyramidal cell dendrites by glutamic acid. Neuropharmacology *13:* 1083–1089 (1964).
25 Duffy, C.; Teyler, T.J.; Shashoua, V.E.: Long-term potentiation in the hippocampal

slice: evidence for stimulated secretion of newly synthesized proteins. Science *212:* 1148–1151 (1981).
26 Duka, T.; Wüster, M.; Schubert, P.; Stoiber, R.; Herz, A.: Selective localization of different types of opiate receptors in hippocampus as revealed by in vitro autoradiography. Brain Res. *205:* 181–186 (1981).
27 Dunwiddie, T.; Mueller, T.A.; Palmer, M.; Stewart, J.; Hoffer, B.: Electrophysiological interactions of enkephalins with neuronal circuitry in the rat hippocampus. I. Effects on pyramidal cell activity. Brain Res. *184:* 311–330 (1980).
28 Finch, D.M.; Babb, T.L.: Demonstration of caudally directed hippocampal efferents in the rat by intracellular injection of horseradish peroxidase. Brain Res. *214:* 405–410 (1981).
29 Fuxe, K.; Hökfelt, T.; Johansson, O.; Jonsson, G.; Lidbrink, P.; Ljungdahl, A.: The origin of the dopamine nerve terminals in limbic and frontal cortex. Evidence for mesocortico-dopamine neurons. Brain Res. *82:* 349–355 (1974).
30 Gähwiler, B.H.: Excitatory action of opioid peptides and opiates on cultured hippocampal pyramidal cells. Brain Res. *194:* 193–203 (1980).
31 Gall, C.; Brecha, N.; Karten, H.J.; Chang, K.-J.: Localization of enkephalin-like immunoreactivity to identified axonal and neuronal populations of the rat hippocampus. J. comp. Neurol. *198:* 335–350 (1981).
32 Goddard, G.V.; McNaughton, B.L.; Douglas, R.M.; Barnes, C.A.: Synaptic change in the limbic system: evidence from studies using electrical stimulation with and without seizure activity; in Livingston, Hornykiewicz, Limbic mechanisms. The continuing evolution of the limbic system concept, pp. 355–368 (Plenum Press, New York 1978).
33 Goodman, R.R.; Snyder, S.H.; Kuhar, M.J.; Young, W.S.: Differentiation of delta and mu opiate receptor localizations by light microscopic autoradiography. Proc. natn. Acad. Sci. USA *77:* 6239–6243 (1980).
34 Haas, H.L.; Ryall, R.W.: Is excitation by enkephalins of hippocampal neurones in the rat due to presynaptic facilitation or to disinhibition? J. Physiol., Lond. *308:* 315–330 (1981).
35 Herkenham, M.; Pert, C.B.: In vitro autoradiography of opiate receptors in rat brain suggests loci of 'opiatergic' pathways. Proc. natn. Acad. Sci. USA *77:* 5532–5536 (1980).
36 Herz, A.; Nacimiento, A.: Über die Wirkung von Pharmaka auf Neurone des Hippocampus nach mikroelektrophoretischer Verabfolgung. Naunyn-Schmiedebergs Arch. exp. Path. Pharmak. *251:* 295–314 (1965).
37 Hökfelt, T.; Jonsson, G.; Ljungdahl, A.: Regional uptake and subcellular localization of [^{3}H]gamma-aminobutyric acid (GABA) in brain slices. Life Sci. *9:* 203–212 (1970).
38 Hökfelt, T.; Ljungdahl, A.; Fuxe, K.; Johansson, O.: Dopamine nerve terminals in the rat limbic cortex: Aspects of the dopamine hypothesis of schizophrenia. Science *184:* 177–179 (1974).
39 Hotson, J.R.; Prince, D.A.; Schwartzkroin, P.A.: Anomalous inward rectification in hippocampal neurons. J. Neurophysiol. *42:* 889–895 (1979).
40 Jung, R.; Kornmüller, A.E.: Eine Methodik der Ableitung lokalisierter Potentialschwankungen aus subcorticalen Hirngebieten. Arch. Psychiat. NervKrankh. *109:* 1–30 (1938).
41 Krnjević, K.; Pumain, R.; Renaud, L.: The mechanism of excitation by acetylcholine in the cerebral cortex. J. Physiol., Lond. *215:* 247–268 (1971).

42 Kromer, L.F.; Björklund, A.; Stenevi, U.: Innervation of embryonic hippocampal implants by regenerating axons of cholinergic septal neurons in the adult rat. Brain Res. *210:* 153–171 (1981).
43 Langmoen, I.A.; Andersen, P.: Summation of excitatory postsynaptic potentials in hippocampal pyramidal cells. J. Neurophysiol. (in press, 1982).
44 Langmoen, I.A.; Segal, M.; Andersen, P.: The mechanism of action of norepinephrine on hippocampal pyramidal cells in vitro. Brain Res. *208:* 349–362 (1981).
45 Lauro, A. di; Schmid, R.W.; Meek, J.L.: Is aspartic acid the neurotransmitter of the perforant pathway? Brain Res. *207:* 476–480 (1981).
46 Lee, H.K.; Dunwiddie, T.; Hoffer, B.: Electrophysiological interactions of enkephalins with neuronal circuitry in the rat hippocampus. II. Effects on interneuron excitability. Brain Res. *184:* 331–342 (1980).
47 Lorente de Nó, R.: Studies on the structure of the cerebral cortex. II. Continuation of the study of the Ammonic system. J. Psychol. Neurol., Lpz. *46:* 113–177 (1934).
48 Lynch, G.S.; Jensen, R.A.; McGaugh, J.L.; Davila, K.; Oliver, M.W.: Effects of enkephalin, morphine and naloxone on the electrical activity of the in vitro hippocampal slice preparation. Expl Neurol. *71:* 527–540 (1981).
49 Lynch, G.S.; Mosko, S.; Parks, T.; Cotman, C.W.: Relocation and hyperdevelopment of the dentate gyrus commissural system after entorhinal lesions in immature rats. Brain Res. *50:* 174–178 (1973).
50 MacLean, P.D.: The limbic system 'visceral brain' and emotional behaviour. Archs Neurol. Psychiat., Chicago *73:* 130–134 (1955).
51 McNaughton, B.L.; Barnes, C.A.; Andersen, P.: Synaptic efficacy and EPSP summation in granule cell of rat fascia dentata studied in vitro. J. Neurophysiol. *46:* 952–966 (1981).
52 Misgeld, U.; Sarvey, J.M.; Klee, M.R.: Heterosynaptic postactivation potentiation in hippocampal CA3 neurons. Long-term changes of the postsynaptic potentials. Exp. Brain Res. *37:* 217–229 (1979).
53 Nadler, J.V.; Vaca, K.W.; White, W.F.; Lynch, G.S.; Cotman, C.W.: Aspartate and glutamate as possible transmitters of excitatory hippocampal afferents. Nature, Lond. *260:* 538–540 (1976).
54 Nadler, J.V.; White, W.F.; Vaca, K.W.; Redburn, D.A.; Cotman, C.W.: Characterization of putative amino acid transmitter release from slices of rat dentate gyrus. J. Neurochem. *29:* 279–290 (1977).
55 Nicoll, R.A.; Alger, B.E.; Jahr, C.E.: Enkephalin blocks inhibitory pathways in the vertebrate CNS. Nature, Lond. *287:* 22–25 (1980).
56 Nicoll, R.A.; Siggins, G.R.; Ling, N.; Bloom, F.E.; Guillemin, R.: Neuronal actions of endorphins and enkephalins among brain regions: a comparative microiontophoretic study. Proc. natn Acad. Sci. USA *74:* 2584–2588 (1977).
57 O'Keefe, J.; Nadel, L.: The hippocampus as a cognitive map (Clarendon Press/Oxford University Press, Oxford 1978).
58 Olton, D.S.; Feustle, W.A.: Hippocampal function required for nonspatial working memory. Exp. Brain Res. *41:* 380–389 (1981).
59 Petsche H.; Stumpf, C.H.; Gogolak, G.: The significance of the rabbit's septum as a relay station between the midbrain and the hippocampus. I. The control of hippocampus arousal activity by the septum cells. Electroenceph. clin. Neurophysiol. *14:* 202–211 (1962).

60 Raisman, G.: Neuronal plasticity in the septal nuclei of the adult rat. Brain Res. *14:* 25–48 (1969).
61 Raisman, G.; Cowan, W.M.; Powell, T.P.S.: An experimental analysis of the efferent projection of the hippocampus. Brain *89:* 83–108 (1966).
62 Rall, W.: Electrophysiology of a dendritic neuron model. Biophys. J. *2:* 145–167 (1962).
63 Rall, W.: Distinguishing theoretical synaptic potentials computed for different soma-dendritic distributions of synaptic input. J. Neurophysiol. *30:* 1138–1168 (1967).
64 Ramon y Cajal, S.: Beiträge zur feineren Anatomie des Grosshirns. I. Über die feinere Struktur des Ammonshornes. Z. wiss. Zool. *56:* 615–663 (1893).
65 Ribak, C.E.; Vaughn, J.E.; Saito, K.: Immunocytochemical localization of glutamic acid decarboxylase in neuronal somata following colchicine inhibition of axonal transport. Brain Res. *140:* 315–332 (1978).
66 Sar, M.W.; Stumpf, W.; Miller, R.; Chang, K.-J.; Cuatrecasas, P.: Immunohistochemical localization of enkephalin in the rat brain and spinal cord. J. comp. Neurol. *182:* 17–38 (1978).
67 Schwartzkroin, P.A.; Andersen, P.: Glutamic acid sensitivity of dendrites in hippocampal slices in vitro. Adv. Neurol. *12:* 45–51 (1975).
68 Scoville, W.B.; Milner, B.: Loss of recent memory after bilateral hippocampal lesions. J. Neurol. Neurosurg. Psychiat. *20:* 11–21 (1957).
69 Segal, M.: The action of serotonin in the rat hippocampal slice preparation. J. Physiol., Lond. *303:* 423–439 (1980).
70 Segal, M.: The action of norepinephrine in the rat hippocampus: intracellular studies in the slice preparation. Brain Res. *206:* 107–128 (1981).
71 Segal, M.; Bloom, F.E.: The action of norepinephrine in the rat hippocampus. I. Iontophoretic studies. Brain Res. *72:* 79–97 (1974).
72 Segal, M.; Bloom, F.E.: The action of norepinephrine in the rat hippocampus. II. Activation of the input pathway. Brain Res. *72:* 99–114 (1974).
73 Skrede, K.K.; Malthe-Sørensen, D.: Increased resting and evoked release of transmitter following repetitive electrical tetanization in hippocampus: a biochemical correlate to longlasting synaptic potentiation. Brain Res. *208:* 436–441 (1981).
74 Skrede, K.K.; Westgaard, R.H.: The transverse hippocampal slice: a well-defined cortical structure maintained in vitro. Brain Res. *35:* 589–593 (1971).
75 Spencer, H.J.; Gribkoff, V.K.; Cotman, C.W.; Lynch, G.S.: GDEE antagonism of iontophoretic amino acid excitations in the intact hippocampus and in the hippocampal slice preparation. Brain Res. *105:* 471–481 (1976).
76 Storm-Mathisen, J.: Glutamate decarboxylase in the rat hippocampal region after lesions of the afferent fibre systems. Evidence that the enzyme is localized in intrinsic neurones. Brain Res. *40:* 215–235 (1972).
77 Storm-Mathisen, J.: High affinity uptake of GABA in presumed GABA-ergic nerve endings in rat brain. Brain Res. *84:* 409–427 (1975).
78 Storm-Mathisen, J.: Tentative localization of glutamergic and aspartergic nerve endings in brain. J. Physiol., Paris *75:* 677–684 (1979).
79 Storm-Mathisen, J.; Fonnum, F.: Quantitative histochemistry of glutamate decarboxylase in the rat hippocampal region. J. Neurochem. *18:* 1105–1111 (1971).
80 Storm-Mathisen, J.; Guldberg, H.C.: 5-Hydroxytryptamine and noradrenaline in the hippocampal region: effect of transection of afferent pathways on endogenous levels,

high affinity uptake and some transmitter-related enzymes. J. Neurochem. *22:* 793–803 (1974).
81 Storm-Mathisen, J.; Iversen, L.L.: Uptake of [³H]glutamic acid in excitatory nerve endings: light and electronmicroscopic observations in the hippocampal formation of the rat. Neuroscience *4:* 1237–1253 (1979).
82 Vanderwolf, C.H.: Neocortical and hippocampal activation in relation to behavior: effect of atropine, eserine, phenothiazines and amphetamine. J. comp. physiol. Psychol. *88:* 300–323 (1975).
83 Vinogradova, D.S.; Brazhnik, E.S.: Neuronal aspects of the septo-hippocampal relations; in Gray, Functions of the septo-hippocampal system. Ciba Foundation Symposium, vol. 58, pp. 145–171 (Elsevier, Amsterdam 1978).
84 Wigström, H.; Swann, J.W.; Andersen, P.: Calcium dependency of synaptic long-lasting potentiation in the hippocampal slice. Acta physiol. scand. *105:* 126–128 (1979).
85 Yamamoto, C.: Long-term potentiation in thin hippocampal sections studied by intracellular recordings. Expl Neurol. *58:* 242–250 (1978).
86 Yamamoto, C.; McIlwain, H.: Electrical activities in thin sections from the mammalian brain maintained in chemically-defined media in vitro. J. Neurochem. *13:* 1333–1343 (1966).
87 Zieglgänsberger, W.; French, E.D.; Siggins, G.R.; Bloom, F.E.: Opioid peptides may excite hippocampal pyramidal neurons by inhibiting adjacent inhibitory interneurons. Science *205:* 415–417 (1979).

P. Andersen, MD, Institute of Neurophysiology, University of Oslo, Karl Johans Gate 47, Oslo 1 (Norway)

Risk Factors for Psychopathology in Individuals with Epilepsy

Janice R. Stevens

National Institutes of Mental Health, Intramural Research Program,
Saint Elizabeth's Hospital, Washington, D.C., and
the University of Oregon Health Science Center, Portland, Oreg., USA

Introduction

Until well into the present century, epilepsy was a disorder of equal interest to psychiatrists and neurologists. This common concern was the result of important differences in the practice of medicine prior to the revolutionary pharmacologic and technical developments of the past 50 years. In that brief period of time, all or nearly all modern anticonvulsants were discovered and the electroencephalogram (EEG) came into use revealing a hitherto unknown world of brain physiology including the cerebral electrical concomitants of the epileptic state. In the last quarter century, study of the biology and chemistry of neurotransmission has exploded into entire new disciplines of neurochemistry, immunohistochemistry, neuropharmacology, psychopharmacology, psychoneuroendocrinology, behavioral neurology, and biological psychiatry, each with their accompanying societies, journals, and contributions. In the midst of this cornucopia of new information, neurology and psychiatry find themselves once again sharing an interest in the pathophysiology and treatment of the same disorders.

The Notion of the 'Epileptic Personality'

Pari passu with the astonishing growth of information concerning nervous system function have come developments in the diagnosis and treatment of epilepsy which have transformed the concept of this disorder from that of a dread and mystifying malady with devastating social and personal consequences to an experimentally replicable derangement of cerebral physiology. Previously studied almost entirely in institutions for the crippled, the retarded, and the insane, it was in this setting of three quarters of a century ago that a new term came into use to describe the untreatable institutionalized patients whose slow-thinking, ponderous, egocentric, aggressive and unctuous intrusiveness were immortalized in the epithet 'epileptic personality'. Not quite immortalized perhaps. For a brief period around the mid-century, neurologists, led by *Lennox* [26], learned that epilepsy per se was not the cause of what might better, but still inexactly, be described as an organic brain syndrome and that this constellation of unpleasant characteristics was rarely observed in patients living outside of institutions.

Psychopathology in Epileptics

Even as the term 'epileptic personality' was being deleted from the medical vocabulary, *Lennox's* former collaborators in Boston, *Erna and Frederick Gibbs* at the Psychiatric Institute of Northwestern University in Chicago published their first startling and exciting report of a very significant increase in psychopathology in a specific group of epileptic patients, namely those with psychomotor seizures [15]. Subsequent studies from their very large series of patients were impressive indeed. Summarized in the second edition of the *Atlas of Electroencephalography* [14], they reported on EEG examinations and clinical correlations of over 10,000 patients, 678 of whom had only psychomotor epilepsy and 1,800 of whom were diagnosed psychomotor and grand mal. Although the methods of clinical diagnosis were not specified, 40% of patients with psychomotor epilepsy and 50% of patients with psychomotor and grand mal seizures were said to have some psychopathology compared to 10% or less for other epilepsies.

Stimulated by their report, I began my own research career with psychomotor epilepsy as the focus, using it as a bridge between the study of the partly known neurology and physiology of the epilepsies and the almost entirely unknown neurology and physiology of the major psychoses. The path was unexpectedly difficult. Despite the collaboration of a number of gifted associates in psychology and psychiatry, and the use of test batteries and standard 'blind' interviews, we were unable to discriminate significant personality characteristics or psychopathology when matched groups of patients with psychomotor epilepsy were compared with age and socio-economically matched patients with clinical and EEG diagnoses of only generalized seizures [35, 42, 46].

Since these results appeared at variance with most of the published literature and with our own expectations, it was important to determine the cause of this discrepancy. The principal factors appeared to be age of, and clinical origin in, patients and control groups [43]. The very large *Gibbs* series did not match psychomotor and generalized epilepsy groups for age. On the contrary, as their own detailed graphs indicate, there is a striking difference in the age of onset of epilepsy in patients with generalized compared with psychomotor epilepsy, and in patients with anterior temporal spike or spike-wave pattern. Since a high proportion of their subjects with grand mal-petit mal were children and the majority of individuals with psychomotor epilepsy were adults, and since the psychopathologies discussed are characteristically of adult onset, it is difficult to compare the relative incidence of psychopathology, in the two groups in a meaningful fashion.

In addition, only 25% of this large sample was diagnosed psychomotor or psychomotor-grand mal epilepsy compared with 50–70% so diagnosed for most adult populations. This further emphasizes the importance of age-matching in a comparison of psychopathology by epilepsy subtype.

A number of anecdotal papers supporting the *Gibbs'* remarkable observation followed the first reports from Chicago [10, 16, 29, 38, 40]. In addition, as *Koella* has reviewed in this volume, work from the animal laboratory clearly indicated that stimulation and lesions of limbic sites led to complex behavior disturbances quite distinct from those which evolved following similar manipulations of the neocortex. Creatively and heuristically gathered together in the well-known work of *MacLean,* the limbic system and its derangements, of which psychomotor epilepsy was but one, became the focus of intense research at the interface of neurophysiology and emotional behavior.

I was fortunate in having the opportunity to work with *MacLean* for 3 years during this fertile period of conceptual and experimental activity. Our investigations were directed to the behavioral effects of cholinergic stimulation of the caudate nucleus of waking unrestrained cats through chronically implanted cannulas. Although not widely appreciated in the present era of catecholamine and peptide popularity, direct instillation of cholinergic agents (mecholyl, carbachol) in micro amounts in the caudate nucleus produces compulsive contraversive circling equal to or surpassing in intensity that produced by dopaminergic stimulation of the same region

[55]. The vast difference between cholinergic activation of the caudate nucleus, the largest component of the neostriatum, and application of similar agents in limbic striatum was illustrated when the injection cannula was placed below the caudate nucleus in the 'motor olfactory striatum' of Papez, i.e., the nucleus accumbens, the largest component of the 'limbic' striatum. Instead of contraversive circling, the animal became alert, vigilant, and began vigorously sniffing and searching about his familiar cage. Similar differences in the effects of stimulation of neopallium and limbic structures at every level are now well known and heighten our expectation that limbic epilepsy – i.e., epilepsy starting or propagating principally in limbic structures, accompanied as it is by a galaxy of subjective emotional and behavioral phenomena reminiscent of all the major psychopathologies, should be associated with an increased incidence of interictal psychiatric disorders as well.

Our laboratory has conducted a number of studies to examine the interictal differences in behavior and personality of individuals with limbic epilepsy compared with generalized and focal nonlimbic seizures. Somewhat to our surprise and contrary to our expectations, none of these studies, whether based on objective or projective psychological tests, blind psychiatric interview or a history of mental hospital admissions demonstrated a significant excess of psychopathology in patients with psychomotor-temporal lobe (limbic) epilepsy (TLE) as reported by the *Gibbs'*. A survey of studies in which matched control populations were used by others supports this conclusion (table I).

Reviewing the published literature concerning epilepsy and psychopathology in 1962, *Tizard* [60] concluded that the data available, largely collected in psychiatric hospitals or neurosurgical clinics, were insufficient to assess a relationship between these disorders. It is remarkable that now, almost 20 years later, published data are still inconclusive. In the sections that follow, complete review of the pertinent literature will not be attempted. Instead, this report will be directed to examination and identification of mechanisms and risk factors which appear to predispose to psychopathology in epilepsy.

Epilepsy and Aggression

A particular association between epilepsy and aggression, and especially between TLE and aggression, has long been assumed. Perhaps this is because every neurologist or psychiatrist has experience of several strik-

Table I. Psychological testing of matched epilepsy and control populations

Investigator	Subjects	Examination	Generalized versus TLE
Mathews and Klove [31]	44 TLE 52 major motor 98 brain-damaged 51 normals	MMPI	no differences
Mignone et al. [34]	98 TLE 53 non-TLE	MMPI	no differences
Glass and Mattson [17]	40 TLE 13 generalized 13 focal/non-TLE	MMPI	no differences
Standage and Fenton [45]	19 TLE 18 non-TLE	present state (Wing)	no differences
Hermann et al. [19]	47 TLE 28 non-TLE	MMPI	no differences
Rodin et al. [39]	A. 56 TLE only B. 22 TLE and generalized C. 46 non-TLE (1 seizure type) D. 32 non-TLE and other	WAIS, MMPI	A. better WAIS, MMPI than B, C, D B. more paranoid personality, psychosis and job problems than A, C, D C. has more organic signs, job difficulty, anticonvulsants than D

WAIS = Wechsler Adult Intelligence Scale. TLE = Temporal lobe epilepsy

ing examples of such cases in a clinical lifetime. Yet, controlled studies consistently fail to support a statistically important relationship between epilepsy and aggressive behavior. As with the studies of other psychopathology, the frequency of aggressive or violent behavior in patients with epilepsy is closely related to the derivation of the population studied. In neurosurgical series, cases of intractable epilepsy or epilepsy plus psychopathology are greatly overrepresented, and centers particularly interested in surgery for aggressive behavior will find a high incidence of such patients referred to them. Other centers less interested in or actually avoiding surgery of patients whose epilepsy is accompanied by psychopathology, report much lower figures.

Examining the records of 700 patients from the Epilepsy Center of Michigan, a less selected group than the patients who are referred for epileptic surgery, *Rodin* et al. [39] found that 34 (4.8%) had a history of destructive, assaultive behavior. The aggressive individuals differed from nonaggressive patients from the same clinic in being largely male, poor, of low IQ, with little religious interest or background, and with more childhood behavior disturbance and signs of diffuse or dominant hemisphere organic pathology. *Mignone* et al. [34] also failed to find a preponderance of aggressive behavior in TLE patients compared with a generalized epilepsy control group studied at the National Institute of Mental Health, Bethesda, Md.

Because epilepsy is more than twice as common in poor socioeconomic groups and deprived racial minorities in US society, and because these social factors are closely associated with aggressive behavior, correlations found between epilepsy and aggressivity in some studies may reflect failure to correct for this factor. Other social predisposers to aggression, including chaotic early environments and brutalization in childhood, show strong correlations with violence and aggression in the nonepileptic population and must be considered in studies of aggressivity in epileptic individuals.

Hermann and Reil [18] have recently attempted this task. After matching patients (from the Mayo Clinic) for age, sex, duration of epilepsy, socioeconomic and other factors they analyzed the Minnesota Multiphasic Personality Inventory (MMPI) score for the *Huesman* et al. [22] agression factor derived from the MMPI (F + Mania + Psychopathic scales). Few differences emerged for TLE and non-TLE on this scale (table II).

The fact that individuals with epilepsy are overrepresented in prison populations has often been cited as evidence of a particular relation between epilepsy and aggression. These data have also seldom been corrected for socioeconomic and IQ factors. Although epilepsy certainly occurs in kings as well as paupers, the importance of socioeconomic factors in predisposing to seizures has only recently been appreciated. If, as these studies have shown, epilepsy, like psychosis and aggressive behavior disor-

Table II. Epilepsy: risk factors for aggression [adapted from ref. 20]

TLE (n = 153) versus generalized epilepsy (n = 79)	p = 0.36 NS
Laterality of EEG spike focus	p = 0.53 NS
Mean chronological age < 30 years	p = 0.001 (GE)
Myoclonic seizures	p = 0.047
Akinetic seizures	p = 0.004

Standard regression analysis excluding variables other than item tested. Aggression score computed by Huesman's aggression scale of MMPI (response conformity plus psychopathic plus hypomania).

ders, is two to three times as common in the poor, significant relationships between epilepsy and other disorders related to poverty and deprivation must be closely examined before causality is assumed.

Aggression and violence in persons with epilepsy, particularly those with TLE, has been extensively investigated via a wide range of assessment methods. In well-matched control studies, clinical history, special rating scales, blind interviews and rating systems, special hostility inventories and studies of prison populations have all failed to show increased aggression in individuals with TLE compared with those with other seizure types [31]. In a striking exception to anecdote and hearsay, *Hermann and Reil* [18] recently reported that aggression is more frequently found in patients with myoclonic and akinetic seizures than any other epileptic diagnostic subgroup (table II).

Epilepsy and Personality Disturbance

A group of Boston investigators has recently rediscovered the once popular, then vigorously rejected concept of the epileptic personality [3, 61]. Hyperreligiosity, 'stickiness', philosophism, hypergraphia, etc. have been attributed by these investigators specifically to patients with TLE. Indeed, the more enthusiastic supporters of this concept go so far as to consider the personality disorders so typical of TLE that the diagnosis can be made clinically from the personality profile, even in the absence of a history of epileptic seizures. Confirmation is achieved by finding spikes

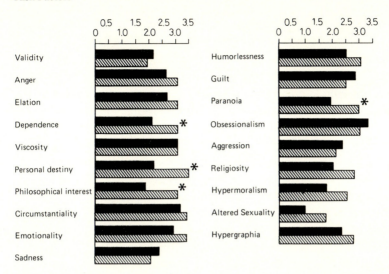

Fig. 1. Replication attempt by *Hermann and Reil* [18] using the *Bear and Fedio* [3] inventory on an age- and socioeconomic-matched sample of patients from an epilepsy clinic. (Only the patients responded to the questionnaire in contrast to the original study in which a family member or other observer also scored each point for the patient). ▧ = Matched TLE (n = 15); ■ = generalized epilepsy (n = 15); *p <0.05.

over the temporal lobe in the waking or sleep EEG [*Geschwind*, personal commun.]. However, epilepsy is not an EEG diagnosis, and the fact that many patients with severe personality disorders or psychoses have abnormal EEGs does not permit a diagnosis of epilepsy. Indeed, the personality traits attributed to TLE are common in psychiatric conditions unaccompanied by epilepsy, in particular manic and posttraumatic psychopathologic states. To demonstrate the specificity of these personality traits for TLE compared with other forms of brain dysfunction, it must be shown that these characteristics are significantly more prevalent in patients with unequivocal psychomotor epilepsy than in other disorders of central nervous system function and in particular other epilepsies. In an investigation which utilized the personality inventory of *Bear and Fedio* to study TLE and non-TLE populations, only four of the 18 features which they characterized as typical of TLE distinguished subjects with TLE from those with generalized epilepsy (fig. 1).

Epilepsy and Psychosis

Of the many publications devoted to this topic, two deserve special attention. In 1952, *Hill* [21] reported that 25–30% of patients diagnosed as schizophrenic had abnormal EEGs. The most common abnormality was spike or slow activity over one or both temporal lobes. Similar observations were subsequently made by others [10, 47, 50, 54]. But, as *Hill* [21] recognized, spikes and slow waves are not seizures and abnormalities over the temporal lobes do not make a diagnosis of TLE unless the patient also has seizures. This logical error has been responsible for much misunderstanding – not only with respect to epilepsy and psychosis but also in relation to abnormal EEGs found in other severe behavior and psychiatric disorders. What the abnormal wave forms over the temporal lobe do mean is that there is an disturbance of brain function within, or projected to, the temporal region [50, 51].

Electroencephalographers privileged to record from multiple intracerebral electrodes require no further evidence of the inadequacy of scalp electroencephalography for localization of epileptic activity of the brain. High voltage epileptiform discharges, particularly those arising in the hippocampal-amygdala complex, are frequently not reflected in the surface EEG. Telemetered EEG studies from chronic implanted electrodes repeatedly demonstrate seizures arising from the side opposite to a scalp slow or spike focus, and even the sphenoid and nasopharyngeal leads are inadequate for localization (fig. 2). While scalp localization of spike or slow activity is not precise enough to allow the designation of the temporal lobe as the source of abnormal activity, with multiple intracerebral electrodes in place, conduction of abnormalities from deep structures to the temporal derivation (as well as vice versa) can be observed [9, 56].

An important and influential paper relating epilepsy and psychosis was published by *Slater* et al. [41] in 1963. These investigators gathered 69 cases of epilepsy and schizophreniform psychosis from the London area, and reasoned on the basis of incidence and prevalence data that the occurrence of these two entities in 69 individuals among a population of nearly 10 million was considerably beyond that to be expected. Ergo, the two ailments must be related. This conclusion is, however, valid only if they had gathered their 69 cases by 'door-to-door' examination of an unselected population. Since, as in their cases, patients were already reporting to hospital for epilepsy (or psychosis) and were then found to also have psychosis (or epilepsy), the coincidence of the two

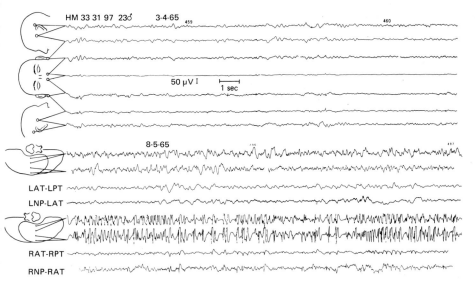

Fig. 2. Interictal scalp, nasopharyngeal and subdural electrode recording from a 23-year-old man with seizures triggered by a picture of a ship or by imagining the image of a ship sailing on water. Seizures are accompanied by genital paresthesias. Note that although the patient is entirely symptom-free at the time of the recording there is continuous epileptiform activity restricted to intracranial subdural electrodes beneath the right temporal lobe. Evidence of this activity is scarcely discernible in the nasopharyngeal recording. RAT, LAT = Right and left anterior temporal derivations; RPT, LPT = Right and left posterior temporal electrodes; RNP, LNP = Right and left nasopharyngeal electrodes.

should not be different from the risk for each ailment in the population at large.

Considerable significance has also been attributed to the mean interval of 14 years between onset of psychosis reported by *Slater* et al. [41]. As *Taylor* [59] has since pointed out, however, a mean interval by definition represents an average, which in the case of psychosis and epilepsy, necessarily emerges from the fact that epilepsy most commonly starts during infancy or childhood while psychoses begin in adults. Thus an interval is necessary between the two and may vary from a few months to 25 years, which, if normally distributed, yields a mean of 14–15 years but 'has no more significance than the mean distance between Boston and other American cities'. Finally, it should be noted that one third of the 69 patients with epilepsy and psychosis of *Slater* et al. [41] were diagnosed

Table III. Psychiatric diagnosis for patients with generalized or psychomotor epilepsy and psychosis (XIIth International Epilepsy Symposium, 1980)

	Cases n	Psychoses	Generalized epilepsy, %	Psychomotor %	Psychomotor and generalized %
Dörr-Zegers and Rauh, Heidelberg	50	brief	76	16	10
		paranoid hallucinatory	7.6	24	70
		schizophreniform	0.5	12	10
Bruens, Breda	57	chronic paranoid syndromes	15	7	54
		schizophreniform	0	2	9.5
Ceard, Milano	94	psychosis with regression	38	50	–
		schizophreniform	18	20.6	–
		depression	28	55	–
Wolf, Berlin	93	'Alt. with productive sx'	23	1.3	23
		paranoid syndromes	8.5	4.7	7.6
		schizophreniform	0	2.3	15.3
		manic depressive	8.5	11.6	15.3

grand mal epilepsy, not TLE, and two thirds of those examined by pneumoencephalogram had evidence of ventricular enlargement or cortical atrophy.

There are, however, other observations which forge a stronger link between epilepsy and psychosis. At the 12th Annual Meeting of the International Epilepsy Congress in Copenhagen in 1980, a special study group met to consider the relationship between epilepsy and psychopathology. Data presented by *Bruens, Ceard, Wolf,* and *Dörr-Zegers and Rauh* [all prepublication] showed that patients with both psychomotor and generalized attacks had an increased incidence of chronic psychosis compared with individuals with 'pure' psychomotor or 'pure' generalized epilepsy (table III). *Rodin* et al. [39] reached a similar conclusion several years previously. They noted that 'pure' psychomotor epilepsy rarely was associated with psychopathology, either on testing or by history, but that patients with psychomotor epilepsy plus generalized seizures had more psychiatric and psychological problems than patients with 'primary' generalized epilepsy or other focal epilepsies accompanied by grand mal (table I).

In what is probably the largest controlled study of TLE patients with and without psychosis, *Kristensen and Sindrup* [24] did not report a difference in frequency of generalized seizures between the two groups, but do not discuss a possible difference in incidence of generalized seizures, nor do they compare prevalence of psychosis in TLE with that of other epilepsies. Items discriminating patients with TLE and psychosis from patients with TLE without psychosis in their study included sinistrality, presence of neurological or radiologic signs of brain damage, bilateral and basal (sphenoid) spike foci on EEG and visceral as opposed to psychic auras. Patients with psychosis had a decreased frequency of psychomotor seizures, an observation also made by *Flor-Henry* [11, 12]. In contrast to both *Flor-Henry* [11, 12] and *Taylor* [59], *Kristensen and Sindrup* [24] did not find that laterality of the epileptic focus or age of onset of seizures discriminated psychotic and nonpsychotic groups.

Most large, and all controlled studies of epilepsy and psychopathology derive from university clinics or public and private institutions. Patients attending university clinics in the United States generally come from lower socioeconomic strata or have particularly intractable forms of epilepsy not amenable to usual treatment. It is thus well to have the information presented by *Zielinski* [62], who found that more than a third of all individuals with epilepsy in Warsaw were not under any medical care at all. These individuals were generally suffering from less severe seizure disorders, usually of the focal (partial) type. If a similar prevalence of untreated individuals with epilepsy obtains elsewhere, the prevalence and incidence statistics for psychopathology in epilepsy, based on samples from populations under medical, surgical, or institutional care, give a skewed picture of the frequency of psychopathology for the majority of individuals with epilepsy.

Neuropathological Studies

Neuropathological studies of individuals with epilepsy are also unfortunately derived largely from institutional settings [28]. In a search of the pertinent literature, I have been unable to find a modern report of complete neuropathological examination of the brain (not just resected temporal lobe) from a series of epileptic patients who were not institutionalized or suffering from severe brain trauma or neoplasm. My own series [53] is similarly flawed but raises some new questions. In the course of studying the pathological material from a series of nearly 100 patients hospitalized and deceased at St. Elizabeth's Hospital, Washington, D.C., with schizophrenic psychoses, histologic material was available for study from 7 patients aged 22–47 years diagnosed psychosis (or schizophrenia) with epilepsy. Histologic sections from a number of brain sites were compared with similar sections from the schizophrenic group (n = 38), with nonschizophrenic neuropsychiatric cases from the same hospital (n = 28), and with a nonpsychiatric nonepileptic control group (18 cases). These examinations indicated that the histological material from patients with epilepsy and psychosis shares with the schizophrenic cases considerable increase in focal fibrillary gliosis with a predilection for periventricular, periaqueductal, midbrain tegmentum and basal forebrain locations (fig. 3).

In addition to sharing the patchy periventricular fibrillary gliosis observed in the schizophrenic material, focal pathology was present in the temporal lobe of 6 patients (hamartoma, horizontal orientation of hippocampal pyramidal cells, vascular anomaly of hippocampus (3 cases). One patient had a grade III astrocytoma of the right temporal lobe invading the basal ganglia and seeding the ependyma. Two of the cases also had severe Purkinje or granule cell loss in the cerebellum. Although the available pathological material was incomplete and did not permit assessment in most cases of homologous structures from both sides of the brain, the observed patchy gliosis affected to a greater or lesser degree the septum, hypothalamus, medial thalamus and midbrain tegmentum, suggesting a healed toxic or infectious process. An illustrative case history follows:

A 22-year-old railway porter had a history of seizures, type not specified, since age 12. Although treated with diphenylhydantoin, he left school at sixth grade because of frequent attacks. There was a history of quick temper which frequently cost him jobs. At age 22 he took to bed, responded poorly to his family, and refused to eat. In 1953, he was transferred to St. Elizabeth's Hospital where a diagnosis of catatonic schizophrenia was made. His EEG was reported as 'diffusely abnormal'. The patient remained remote and uncommunicative, preoccupied with auditory hallucinations. Twice during the year in the hospital he was found on the floor unconscious, but seizures were not witnessed. He expired 1 year after admission following a thoracotomy and pulmonary lobectomy for tuberculosis.

Neuropathological examination (by *M. Neuman*) revealed a grossly normal brain weighing 1,350 g. Microscopic examination demonstrated that pyramidal cells in the right hippocampus were oriented horizontally. There was severe Purkinje cell loss in the cerebellar cortex. Healed ependymal granulations were present as well as a heavy infiltration of corpora amylacea in the vallecula region (junction of lateral substantia innominata and amygdala). There was moderate diffuse gliosis of the hippocampus and a thin-walled vascular anomaly beneath Somer's sector. There was also mild diffuse fibrillary astrocytosis in globus pallidus, ansa lenticularis, anterior perforated substance, periventricular nucleus of hypothalamus and lateral geniculate nucleus.

Fig. 3. Brain section from a 23-year-old hospitalized patient with a schizophrenic psychosis and with a history of epilepsy, type undetermined. Holzer stain for glial fibrils. Unmagnified whole brain section at the level of the septal nuclei. Darker area surrounding lateral and third ventricle represents dense fibrillary gliosis.

Comment. This young man, who entered the hospital in a state diagnosed as catatonia and with a history of generalized seizures, had a unilateral hippocampal anomaly which may have served as the seizure focus. In addition, there was evidence of diffuse gliosis in subcortical nuclei similar to that seen in schizophrenic brains [36, 52] and suggestive of a toxic or infectious process. Similar gliosis in epileptic patients has been ascribed to anoxic encephalopathy secondary to ictus. Since similar changes occur in paramedial subcortical nuclei without preceding epilepsy, it is equally possible that both the major psychoses and the psychoses occurring in some patients with epilepsy are due to the same underlying process, e.g., an infectious, autoimmune or toxic encephalopathy.

Further evidence that deep and midline pathologic changes predispose to psychosis emerges from *Malamud's* [27] study of tumors of the limbic system presenting with major psychoses. Of 18 patients with 'limbic

Table IV. Psychopathology with tumors of limbic system [from ref. 27]

Site	Cases n	Seizures	'Schizophrenic'	Depression	Mania	Neurosis
Temporal lobe	9	9	4 (2L, 2R)	4 (2L, 2R)		
IIIrd ventricle	7	1	4		1	2

tumors' hospitalized and deceased in a California state mental hospital or university psychiatric service, 10 were diagnosed schizophrenia (table IV). *Malamud's* [27] study suggests that it is the site and nature of the pathology that is the critical factor responsible for the psychosis and not the presence or absence of seizures. On the basis of EEG, radiologic and clinical data, *Kristensen and Sindrup* [24] reached similar conclusions.

Antithesis between Epilepsy and Psychosis?

The observation that seizures and psychosis may be in some fashion antithetical to each other led to the introduction of convulsive therapy for the major psychoses. This observation was subsequently supported by *Landolt* [25], who proposed the term 'forced normalization' of the EEG, accompanied by suppression of ictus as an important precipitant of psychosis in epilepsy. A decrease in psychomotor seizure incidence in patients with TLE who develop psychosis was subsequently documented by *Flor-Henry* [11–13] and by *Kristensen and Sindrup* [24]. Onset of psychosis months or years following relief of temporal lobe epilepsy by temporal lobectomy has been documented by *Jensen and Larsen* [23], *Taylor* [58] and *Stevens* [46].

Biochemical Factors

Although the many similarities between subjective symptoms and stereotyped automatisms of schizophrenia and psychomotor epilepsies suggest a related anatomy [47, 48], pharmacology provides clues that psychotic behaviors are mediated by systems which differ biochemically from the substrates responsible for initiating or maintaining epileptic seizures. With few exceptions, pathologic violence, paranoia and hallucinations respond more favorably to catecholamine-blocking neuroleptics or to lithium carbonate than to anticonvulsant agents. Yet neuroleptic agents and other means of effecting catecholamine blockade increase the propensity for epileptic seizures in man and in many animal models of epilepsy [7, 57]. Conversely, catecholamine precursors and agonists generally increase seizure threshold [1]. Additionally, recent evidence indicates that generalized seizures induce subsensitivity in specific central catecholamine binding sites [4, 6] and increased sensitivity in others [33]. These facts suggest that the catecholamine system may be part of the brain's natural defense against propagation of focal epileptic discharge. This surmise is supported by the increased risk of psychosis when seizures, especially psychomotor seizures, decrease in frequency.

It has been suggested elsewhere that highly localized subcortical seizure-like discharges recorded from hypothalamus and brain stem during sleep, sexual orgasm, lactation and ovulation of normal mammals may represent imperative biologic signals which utilize strictly localized 'excessive neuronal discharge' (*H. Jackson's* definition of seizures) or hypersynchrony to transmit essential life- and species-preserving information over the background noise of CNS activity [50, 51]. Such focal discharges, associated with vital functions for survival, must be restricted to pertinent areas by steep voltage gradients to protect the organism from their propagation to centers of consciousness and motor behavior. Evidence that petit mal epilepsy is antagonized by naloxone [44] suggests that endogenous opioids have a role in 'physiologic' seizure generation and that petit mal may represent failure of focal 'miniseizure' containment.

It is likely that the GABA and catecholamine systems, both powerful and ubiquitously dispersed inhibitors of neuronal discharge, normally participate in restricting these focal discharges to physiologically relevant subcortical regions. The remarkable rise in the incidence of epilepsy, psychosis, and pathologic violence associated with puberty suggests that an increase in both the propensity for focal subcortical epileptiform discharge and in the physiologic resistance to its propagation are associated with the cerebral events accompanying the onset of sexual activity.

Both amygdala and hippocampus are intimately involved with the events preceding initiation of reproductive life. Observations by Taylor [59] and others associating psychoses of epileptic patients with pubertal onset of seizures (fig. 4) may provide an important clue to mechanisms

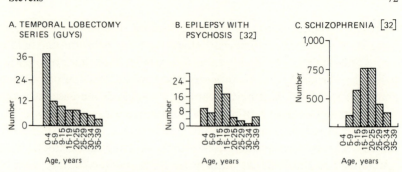

Fig. 4. Comparison of age of onset of different groups of epileptic patients. *A* Group derived from Guys Hospital neurosurgical series and largely free of psychopathology. *B* Group derived from *Slater* et al. [41]: series of 69 patients with schizophreniform psychoses and epilepsy; (two thirds TLE; one third generalized epilepsy). *C* Age of first admission to hospital of schizophrenic patients in Great Britain. The similar age of onset of epilepsy subsequently associated with psychosis to age of onset of uncomplicated schizophrenia suggests that the schizophrenic psychoses occurring in individuals with epilepsy are due to similar predisposing events as schizophrenic psychoses without epilepsy [32].

regulating seizure threshold in this normally seizure-prone region of the mammalian brain.

Pharmacologic evidence suggests that paranoia, hallucinations and pathologic violence are maintained by, or depend upon, circuitry that differs from and even opposes the mechanisms maintaining or propagating epileptic discharge. Further evidence of antagonism between seizures and psychosis emerges when anticonvulsant drugs are added to the neuroleptic regimen of patients with chronic schizophrenia (but no history of epilepsy) who have responded only partially or poorly to neuroleptic treatment. 12 such patients have been treated with anticonvulsant medication by the author. All were institutionalized and were insufficiently improved, even by massive doses of several different neuroleptics, to leave the hospital. While maintained on their usual neuroleptic regimens, 10 were treated with carbamazepine, 9 with sodium valproate, 3 with acetazolamide and 1 with diphenylhydantoin. 9 of the 10 schizophrenic patients grew worse during treatment with carbamazepine. Two developed severe dystonic reactions relieved by intravenous diphenhydramine. 1 patient, hospitalized with a schizoaffective disorder for many years, was moderately improved on carbamazepine. This man had also responded well to lithium carbonate, which, however, induced excessive water drinking followed by seizures. This patient's psychosis was subsequently even more improved by treatment with sodium valproate and neuroleptics, and he is now in the process of discharge from the hospital after 26 years of confinement. 3 other patients were modestly improved on sodium valproate but 1 patient, diagnosed paranoid schizophrenia, developed a psychotic delirium while on the drug (this patient was also receiving lithium and haloperidol). Of 5 patients treated with acetazolamide, none had a significant change in behavior.

In contrast to the consistently unfavorable effect of anticonvulsants in schizophrenic psychoses, *Okuma* et al. [37] and *Ballenger and Post* [2] have reported good results using carbamazepine for both the treatment and prevention of recurrent illness in subgroups of manic-depressive patients. These, and additional results are presented by *Post* in this volume. *Emrich* et al. [8] had similar success with sodium valproate. These discrepancies in response of schizophrenic and manic-depressive patients to carbamazepine and valproate are among the very few pharmacotherapeutic distinctions between schizophrenia and mania.

Risk Factors Predisposing to Psychosis in Epilepsy

Many reports indicate that patients with epilepsy as a group have a higher incidence of psychopathology than control populations, but even this common assumption has been challenged when controls were selected from matched general medical or nonepileptic neurologic populations. The incidence of psychopathology is much higher in speciality epilepsy and neurosurgical clinics or in units serving a disproportionately high number of unemployed or disabled individuals. Since TLE, with or without generalized seizures, accounts for approximately 70% of all adult epilepsy [30], it is not surprising that adults with both epilepsy and psychopathology usually have TLE. The interesting and critical question is to identify the factors that predispose only certain patients with TLE to psychopathology.

Clinical Factors Predisposing to Psychosis in Epilepsy

From this brief survey of clinical and pathological material, a number of risk factors which are positively or negatively associated with psychopathology in epilepsy in general and TLE in particular can be identified.

Risk Factors for Psychopathology and Epilepsy in General

Seizures, like headache, are not a disease but a symptom. The incidence of psychopathology accompanying the epilepsy appears to depend to a large extent on the underlying cause of the seizures. When epilepsy is a symptom of a severe untreatable metabolic, morphologic or physiologic defect, the incidence of associated psychopathology is usually high. When epilepsy is a symptom of a morphologically and physiologically intact brain (insofar as this can be ascertained by radiologic and electrophysiologic study), the incidence of psychopathology is much reduced. New diagnostic techniques such as the PET scan, which reveals heretofore unknown information concerning hyper- and hypometabolic areas associated with the epileptic focus even during the interictal period, give important new information on previously unknown aspects of cerebral malfunction. According to their extent and position, these focal areas of altered metabolism are likely to contribute to interictal symptoms of patients with epilepsy. Low IQ, social and interpersonal deprivation and family history of mental disease have also been associated with increased risk of psychopathology in patients with seizures [5].

Risk Factors for Psychopathology in Patients with TLE

These have been much more thoroughly studied. Comparing a large series of TLE patients with and without psychosis, *Kristensen and Sindrup* [24] found a higher incidence of automatisms and epigastric auras, fewer psychic auras, more widespread EEG abnormality, including multifocal and basal (sphenoid) spike activity, a higher incidence of abnormal neurologic signs and cognitive impairment and of diffuse ventricular enlargement in patients with psychoses. Although side of focus and age at onset did not distinguish psychotic and nonpsychotic groups, there were significantly more sinistrals in the psychotic group. As the vast majority of the population is right-handed, they rightly note that the preponderance of sinistrals in the psychotic group may signify brain damage in early life but does not necessarily indicate a predominance of lateralized left hemisphere pathology. As in the study of *Rodin* et al. [39], their patients with psychopathology more often experienced onset of epilepsy in relation to a CNS infection. They concluded that epileptic psychoses are not due to epilepsy per se, but are organic psychoses due to structural damage to deep limbic areas of the brain and that such damage is responsible for both the epilepsy and the psychosis.

Comparison of data from large, well-matched groups of epilepsy

Risk Factors

Table V. Risk factors for psychopathology in patients with epilepsy

	A, %	B, %	A/B
Psychomotor epilepsy [34]	30.6	30.2	1.0 ×
TLE [46]	30.0	30.0	1.0 ×
TLE (MMPI) [19]	22.0	30.0	0.73 ×
Age of onset 9 years[1]	22.3	5.0	4.5 ×
Sphenoid spikes[2]	21.0	5.0	4.2 ×
Multiple spikes[2]	22.7	6.0	3.8 ×
Automatism or visceral aural[2]	11.4	6.4	1.8 ×
Abnormal neurological examination	21.0	8.3	2.5 ×
Psychomotor seizure frequency 1/week[1]	6.1	11.6	0.525 ×
History of febrile seizures[1]	6.0	9.5	0.63 ×
Family history of epilepsy[1]	5.2	10.6	0.43 ×

A = Number of patients with psychosis, psychiatric hospitalizations, or other severe psychopathology (PS) with risk factors (RF), divided by total PS with RF; B = number with PS without RF, divided by total PS without RF; A/B = increase or decrease of PS in risk with each RF.
[1] From *Kristensen and Sindrup* [24]. Arbitrary incidence of psychosis: 10%.
[2] From *Taylor* [58.] Arbitrary incidence of schizophrenia-like psychosis: 10%.

patients with and without psychopathology allows a more precise statement of the risk factors which predispose to, or protect individuals with TLE from, psychopathology. If the large groups of patients studied by *Kristensen and Sindrup* in Odense and by *Taylor* in London can be considered typical of TLE populations with and without psychosis, and if the prevalence of psychosis in TLE patients is known or arbitrarily assigned, the attributes studied in their psychotic and nonpsychotic TLE patients can be examined as risk factors. As is evident from table V, of the factors reported to date, basal EEG spiking and evidence of diffuse anatomic and physiologic disturbance make the greatest contribution to probability of psychosis in these populations. Examination of the interaction of these variables, and of other factors, including historical and psychosocial data, should allow more precise definiton of the risk and determinants of psychopathology in epilepsy.

For the past 25 years, the old pejoratives formerly encompassed in the term 'epileptic personality' have been applied to individuals with TLE. Vivid anecdotes and faulty statistics added violence and schizophreniform psychoses to the burden already borne by individuals with seizures. Yet, study of individuals with epilepsy living and working in the community, perhaps one third of whom are without medical treatment, of patients attended by private physicians, and of carefully matched socioeconomic groups indicates that the risk of these psychopathological disorders has been exaggerated. This is at least in part because most of the published literature on this subject is from psychiatric research and neurosurgical centers specializing in the most complex and intractable forms of epilepsy. The importance of TLE in predisposing to psychopathology remains an unanswered question. Well-controlled clinical studies to date have not shown that TLE is the most significant variable predisposing patients with epilepsy to psychiatric disorders. The most powerful neurobiological factors identified to date are clinical, EEG, radiologic and histologic evidence of bilateral deep or diffuse cerebral pathology or pathophysiology.

Conclusions

Epilepsy, including temporal lobe epilepsy, is a symptom of underlying cerebral dysfunction and probably per se is not or is only rarely a cause of chronic psychopathology. The latter, like the epileptic seizure is rather another manifestation of cerebral dysfunction. Epileptic seizures originating in or propagating to the limbic system produce signs and symptoms which mimic not only many normally experienced emotions, but pathologic emotional states, perceptions and behaviors which closely resemble experiences in the major psychoses. In a minority of affected individuals, epilepsy is associated with chronic psychopathology. The proportion of individuals affected is strongly influenced by the population studied and will vary from 100% in psychiatric or neurobehavior clinics to around 25% in university clinics in the United States, and to less than 10% in private practices for employed individuals [49]. Whether there is a particular relationship between psychomotor-temporal lobe (limbic) epilepsy and psychopathology compared with other epilepsies is still difficult to conclude from the data at hand. However, it appears that the patients with 'pure' TLE and no other form of epilepsy are not at higher risk for psychopathology than patients with generalized epilepsy. Principal risk factors identified for psychosis are evidence of deep or diffuse structural abnormality of the brain, basal or multiple spike foci, postchildhood onset of seizures, and abnormal neurologic examination. Risk factors for aggressive behavior include low IQ, poor socioeconomic environment, early age of onset and akinetic seizures. Neuropathologic studies suggest that schizophrenia, like psychoses associated with epilepsy, results from still unidentified toxic, infectious anoxic or traumatic factors which appear to cause the neuropathologic changes found in psychosis with or without epilepsy.

In this review an attempt has been made to discriminate the important risk factors which predispose some patients with epilepsy to psychosis. By identifying the variables predisposing the individual with epilepsy to psychopathology, we may be in a stronger position to protect or treat those at high risk and to seek the fundamental pathology which leads to phenomenologically similar disturbances of brain function.

References

1 Anlezark, G.M.; Horton, R.W.; Meldrum, B.S.: Dopamine agonists and reflex epilepsy. Adv. biochem. Psychopharmacol. *19:* 383–388 (1978).
2 Ballenger, J.C.; Post, R.M.: Therapeutic effects of carbamazepine on affective illness: a preliminary report. Commun. Psychopharmacol. *2:* 159–175 (1978).
3 Bear, D.M.; Fedio, P.: Quantitative analysis of interictal behavior in temporal lobe epilepsy. Archs Neurol. *34:* 454–467 (1977).
4 Bergstrom, D.A.; Kellar, K.J.: Effects of electroconvulsive shock on monoaminergic receptor binding sites in rat brain. Nature, Lond. *278:* 464–465 (1979).
5 Blumer, D.: Treatment of patients with seizure disorder referred because of psychiatric complications; in Blumer, Levin, Psychiatric complications in the epilepsies: current research and treatment. MacLean Hosp. J., special issue, pp. 53–73 (1977).
6 Chiodo, A.; Antelman, S.M.: Electroconvulsive shock. Progressive dopamine autoreceptor subsensitivity independent of repeated treatment. Science *210:* 799–801 (1980).
7 Corcoran, M.E.; Fibiger, H.C.; McCoughran, J.A., Jr.; Wada, J.A.: Potentiation of amygdaloid kindling and metrazol induced seizures by 6-hydroxydopamine in rat. Expl. Neurol. *45:* 118–133 (1974).
8 Emrich, H.; Zerssen, D.V.; Altmann, H.; Kissling, W.; Moller, H.J.: Effect of sodium valproate on mania. Arch. Psychiat. NervKrankh. *229:* 1–16 (1980).
9 Engel, J., Jr.; Rausch, R.; Lieb, J.P.; Kuhl, D.E.; Crandall, P.H.: Correlation of criteria used for localizing epileptic foci in patients considered for surgical therapy of epilepsy. Ann. Neurol. *9:* 215–224 (1981).
10 Ervin, F.; Epstein, A.W.; King, H.E.: Behavior of epileptic and nonepileptic patients with 'temporal spikes'. Archs Neurol. Psychiat. *74:* 488–497 (1955).
11 Flor-Henry, P.: Psychosis and temporal lobe epilepsy. Epilepsia *10:* 363–395 (1969).
12 Flor-Henry, P.: Ictal and interictal psychiatric manifestations of epilepsy. Specific or non-specific? Epilepsia *13:* 773–783 (1972).
13 Flor-Henry, P.: Epilepsy and psychopathology; in Granville-Grossmann, Recent advances in clinical psychiatry (Churchill, Livingston, Edinburgh 1976).

14 Gibbs, F.A.; Gibbs, E.L.: Atlas of electroencephalography; vol. 2, 3 (Addison-Wesley Press, Cambridge, Mass. 1964).
15 Gibbs, F.A.; Gibbs, E.L.; Furster, B.: Psychomotor epilepsy. Archs Neurol. Psychiat. 60: 331–339 (1948).
16 Glaser, G.H.: The problems of psychosis in psychomotor temporal lobe epileptics. Epilepsia 5: 271–278 (1964).
17 Glass, D.H.; Mattson, R.H.: Psychopathology and emotional precipitation of seizures in temporal lobe epileptics. Proc. 81th Annu. Convention Am. Psychol. Ass., vol. 8, pp. 425 (1973).
18 Hermann, B.P.; Reil, P.: Interictal personality correlates of temporal lobe and primary generalized epilepsy. Cortex 17: 125–128 (1981).
19 Hermann, B.P.; Schwartz, M.S.; Karnes, W.; Vandar, P.: Seizure type and age of onset: relationship to psychopathology in epilepsy. Epilepsia 21: 15–23 (1980).
20 Hermann, B.P.; Schwartz, M.S.; Whitmar, S.; Karnes, W.: Aggression in epilepsy: seizure type comparisons and high risk variables (submitted, 1981).
21 Hill, D.: EEG in episodic psychotic and psychopathic behavior. Electroenceph. clin. Neurophysiol. 4: 419–442 (1952).
22 Huesman, L.R.; Lefkowits, M.M.; Eron, L.R.: Sum of MMPI scales F, 4 and 9 as a measure of aggression. J. consult. clin. Psychol., 40: 1071–1079 (1978).
23 Jensen, I.; Larsen, J.K.: Mental aspects of temporal lobe epilepsy. Follow-up of 74 patients after resection of a temporal lobe. J. Neurol. Neurosurg. Psychiat. 42: 256–265 (1979).
24 Kristensen, O.; Sindrup, E.H.: Psychomotor epilepsy and psychosis. Acta neur. scand. 57: 361–377 (1978).
25 Landolt, A.: Serial EEG investigations during psychotic episodes in epileptic patients and during schizophrenic attacks, in Lorenz de Maas, Lectures on epilepsy, pp. 91–133 (Amsterdam, 1958).
26 Lennox, W. G.: Epilepsy and related disorders, vol. I, II (Little, Brown, Boston 1960).
27 Malamud, N.: Psychiatric disorder with intracranial tumors of limbic system. Archs Neurol. 17: 113–123 (1967).
28 Margerison, J.H.; Corsellis, J.A.N.: Epilepsy and the temporal lobes. Brain 89: 499–529 (1966).
29 Mark, V.H.; Ervin, F.R.: Violence and the brain (Harper & Row, New York 1970).
30 Marques-Assis, L.: Discussion; in Birkmayer Epileptic seizures, behaviour, pain, pp. 89–90 (University Park Press, Baltimore 1976).
31 Matthews, C.G.; Klove, H.: MMPI performances in major motor, psychomotor, and mixed seizure classifications of known and unknown etiology. Epilepsia 9: 43–53 (1968).
32 Mayer-Gross, W.: Clinical psychiatry (Williams & Wilkins, Baltimore 1960).
33 McWilliam, J.R.; Meldrum, B.S.; Checkley, S.A.: Enhanced growth hormone response to clonidine after repeated electroconvulsive shock in a primate species. Psychoneuroendocrinology 6: 77–79 (1981).
34 Mignone, R.J.; Donnelly, E.F.; Sadowsky, D.: Psychological and neurological comparisons of psychomotor and non-psychomotor epileptic patients. Epilepsia 11: 345–359 (1970).

35 Mirsky, A.F.; Primac, D.W.; Marsan, C.A.; Rosvold, H.E.; Stevens, J.R.: A comparison of the psychological test performance of patients with focal and non-focal epilepsy. Expl Neurol. *2:* 75–89 (1960).
36 Nieto, D.; Escobar, A.: Major psychoses; in Minckler, Pathology of the nervous system (McGraw-Hill, New York 1972).
37 Okuma, T.; Kishimoto, A.; Inoue, K.; Matsumoto, H.; Ogura, A.; Matsushita, T.; Naklao, T.; Ogura, C.: Anti-manic and prophylactic effects of carbamazepine on manic-depressive psychosis. Folia psychiat. neurol. jap. *27:* 283–297 (1973).
38 Pond, D.A.: Psychiatric aspects of epilepsy. J. Indian med. Prof. *3:* 1441–1451 (1957).
39 Rodin, E.A.; Katz, M.; Lennox, K.: Differences between patients with temporal lobe seizures and those with other forms of epileptic attacks. Epilepsia *17:* 313–320 (1976).
40 Serafetinides, E.A.: Aggressiveness in temporal lobe epilepsy and its relation to cerebral dysfunction and environmental factors. Epilepsia *6:* 33–47 (1965).
41 Slater, F.; Beard, A.W.; Glithero, E.: The schizophrenia-like psychoses of epilepsy. Br. J. Psychiat. *109:* 95–150 (1963).
42 Small, J.; Milstein, V.; Stevens, J.R.: Are psychomotor epileptics different? A controlled study. Archs Neurol. *7:* 187–194 (1962).
43 Small, J.G.; Small, I.F.; Hayden, M.P.: Further psychiatric investigations of patients with temporal and nontemporal lobe epilepsy. Am. J. Psychiat. *123:* 303–310 (1966).
44 Snead, O.C.; Bearden, L.F.: Anticonvulsants specific for petit-mal antagonize epileptogenic effect of leucine enkephalin. Science *210:* 1031–1032 (1980).
45 Standage, K.F.; Fenton, G.W.: Psychiatric symptom profiles of patients with epilepsy: a controlled investigation. Psychol. Med. *5:* 152–160 (1975).
46 Stevens, J.R.: Psychiatric implications of psychomotor epilepsy. Archs gen. Psychiat. *14:* 461–472 (1966).
47 Stevens, J.R.: Psychomotor epilepsy and schizophrenia; in Brazier, Epilepsy: its phenomena in man. UCLA Forum in Medical Sciences (Academic Press, New York 1973).
48 Stevens, J.R.: An anatomy of schizophrenia? Archs gen. Psychiat. *29:* 177–189 (1973).
49 Stevens, J.R.: Interictal clinical manifestations; in Penry, Daly, Complex partial seizures (Raven Press, New York 1975).
50 Stevens, J.R.: The EEG spike: signal of information transmission? A hypothesis. Ann. Neurol. *1:* 309–314 (1977).
51 Stevens, J.R.: All that spikes is not fits; in Shagass, Gershon, Friedhoff, Psychopathology and brain dysfunction, pp. 183–198 (Haven Press, New York 1977).
52 Stevens, J.R.: Biologic background of psychoses in epilepsy; in Canger, Angeleri, Penry, Advances in epileptology, pp. 167–172 (Raven Press, New York 1980).
53 Stevens, J.R.: Neurology and neuropathology of schizophrenia; in Nasrallah, Henn, Schizophrenia as a brain disease (University Press, New York 1981).
54 Stevens, J.R.; Bigelow, L.; Denney, D.; Lipkin, J.; Livermore, A.; Rauscher, F.; Wyatt, R.J.: Telemetered EEG-EOG during psychotic behaviors of schizophrenia. Archs gen. Psychiat. *36:* 251–262 (1979).
55 Stevens, J.R.; Kim, C.; MacLean, P.D.: Stimulation of caudate nucleus: behavioral effects of chemical and electrical excitation. Archs Neurol. *4:* 47–54 (1961).
56 Stevens, J.R.; Mark, V.H.; Ervin, F.; Pacheco, P.; Suematsu, K.: Deep temporal stimulation in man. Archs Neurol. *21:* 157–169 (1969).
57 Tabakaff, B.; Yanai, J.; Ritzman, R.F.: Noradrenaline and seizures. Science *203:* 1265–1266 (1978).

58 Taylor, D.C.: Mental state and temporal lobe epilepsy. A correlative account of 100 patients treated surgically. Epilepsia *13:* 727–765 (1972).
59 Taylor, D.C.: Epileptic experience, schizophrenia and the temporal lobe. McLean Hosp. J., p. 21 (1977).
60 Tizard, B.: The personality of epileptics. A discussion of evidence. Psychology *59:* 196–202 (1962).
61 Waxman, S.G.; Geschwind, N.: Hypergraphia in temporal lobe epilepsy. Neurology *24:* 629–636 (1974).
62 Zielinski, J.J.: Epidemiology of medico-social problems of epilepsy in Warsaw (DHEW, Washington, DC 1974).

J.R. Stevens, MD, National Institutes of Mental Health, Intramural Research Program, Saint Elizabeth's Hospital, Washington, DC 20032 and University of Oregon Health Science Center, Portland, OG 97201 (USA)

The Effect of the Location of an Epileptogenic Lesion on the Occurrence of Psychosis in Epilepsy

Ira Sherwin[1]

Department of Neurology, Harvard Medical School, Boston, Mass., USA

Introduction

The tendency for psychiatric abnormalities to develop in association with epilepsy has been recognized since ancient times [32]. At the turn of this century *Gowers'* [6] *Borderland of Epilepsy* and his subsequent works [7] presaged the specificity of this concept with remarkable insight. Based on these essays and numerous case reports that have followed, it does not seem overly extravagant to suggest that nowhere do the disciplines of neurology and psychiatry form a more widely overlapping borderland than in the case of epilepsy.

More recently, however, partly as a result of having demonstrated an underlying disturbance in cerebral electrogenesis and partly as a result of well intentioned efforts to destigmatize this disorder, a continuing effort began seeking to repudiate and deny the notion of this association [14, 18].

In what follows I will attempt to show that this notion of nonspecificity is not warranted by the available facts. That notwithstanding, one serious consequence of such assertions has been that neurologists accepting this point of view have tended to attribute psychiatric symptoms in their epileptic patients to *nonspecific* situational reactions or worse still have ignored them completely. Psychiatrists on the other hand frequently see these patients as having 'organic brain disease' and, therefore, outside their

---

[1] The author is pleased to acknowledge his indebtedness to his colleagues at the Neurology/Neurosurgical Units of the Boston City Hospital, Reed Neurological Research Center, UCLA, and INSERM (U97), Paris, where these studies were performed and without whose assistance they could not have been undertaken.

area of concern, and in some cases, expertise. The result has been that this borderland of disciplines has been transformed into a no-man's-land in which such patients receive only incomplete treatment for serious and complex problems. Just how widespread this problem actually is, is difficult to say. In a recent report by the Commission for the Control of Epilepsy [16], it was noted that '12% of the 200,000 persons in institutions for the mentally ill have epilepsy'. This is a remarkable finding in and of itself, because it is commonly estimated that the prevalence of epilepsy in the general population (USA) is only about 1% [1]. Additionally, however, when it is realized that these hospitalized patients represent only the most serious cases, it is probable that in terms of the actual prevalence of psychopathology associated with epilepsy, this statistic is extremely conservative.

Based on original observations as well as those of others, it is our position that there exists not only a *generally* increased prevalence, but that there is a high degree of *specificity* between the type of epilepsy and the type of associated psychopathology. It is only through an increased awareness of the details of this association that the physician responsible for such patients can begin to provide them the comprehensive care they require.

Our Own Studies

In a study in which patients were selected for the concordance of medically intractable epilepsy and behavior disorder [24], we noted that 15% of the patients had demonstrated a schizophrenia-like psychosis at some point in their illness. Based on pneumoencephalographic and electroencephalographic (EEG) findings, all of the psychotic patients appeared to have left-sided temporal lobe epileptogenic lesions.

Although this finding, i.e. that patients with left temporal lobe epileptogenic lesions are peculiarly vulnerable to develop a schizophrenia-like psychosis, was consistent with other reports [3], the small number of cases dictated that a larger series be studied. In doing so, a major difficulty encountered in attempting to establish the specificity of the relationship of specific forms of psychopathology to the laterality of epileptogenic lesions derived from the fact that, in seeking to define the laterality of an epileptogenic lesion, EEG findings may be grossly misleading. When *interictal* data alone are relied on to lateralize the epileptogenic lesion, even the use of nasopharyngeal, sphenoidal or depth electrodes does not eliminate the

Table I. Series composition (n = 112)

Category	Number of Patients
Lost to follow-up or dead	13
Cases remaining for analysis	99
Temporal lobe epilepsy	90
Unilateral (lobectomy)	63
Bilateral (depth electrode only)	27
Other epilepsies	9

problem [5]. Therefore, we sought to determine the relative frequency of psychoses associated with left-or right-sided temporal lobe epileptogenic lesions in a group of patients in whom the laterality was established with sufficient certainty, based on depth recorded *ictal* episodes, to justify a temporal lobectomy.

The data constituting this second study were derived from a records review of the UCLA series which extends over a decade, and which at the time of analysis encompassed 112 patients.

Table I shows the composition of the entire series. Of this series, only the cases (n=63) with clear unilateral lesions warranting temporal lobectomy were further analyzed. There were 36 males and 27 females in this group. Left temporal lobectomies were performed in 17 cases and right temporal lobectomies in the remaining 46. The predominance of patients with right-sided lobectomies in this sample of 63 cases is highly significant ($\chi^2 = 13.65$, $p < 0.001$), and is a point to which we shall return.

In table II the distribution of the laterality of the epileptogenic lesions by sex is shown. As can be seen, left- and right-sided foci tend to be unequally distributed for the sexes (Fisher Exact Probability = 0.053). The Binomial Test (assuming no a priori laterality bias) reveals that for males, the left/right difference is not significant. For females, however, this difference is highly significant, $p < 0.001$. This relative underrepresentation of females with left-sided foci is of special interest since *Taylor* [30] has postulated that this is the group most at risk for developing the schizophrenia-like psychosis associated with temporal lobe epilepsy. Despite the prior attempt to exclude patients with active psychopathology from the UCLA series, we found 7 cases with unilateral epileptogenic lesions warranting temporal lobectomy, who were diagnosed as having been psychotic.

Table II. Temporal lobectomy group: laterality and sex (n = 63)

Sex	Left (n = 17)	Right (n = 46)
Male	13	23
Female	4	23

Table III. Psychosis and laterality (n = 61)

Lobectomy	Psychotic (n = 7)	Nonpsychotic (n = 54)
Right	2	44
Left	5	10

Of these 7 cases, 6 were diagnosed as having had a paranoid or schizophrenic type of psychosis. Of the 6, one had undergone a right and the rest a left temporal lobectomy. The 7th patient (with a right temporal lobectomy) was diagnosed as having had a schizo-affective psychosis with marked depressive features. The relative frequencies of left- and right-sided foci in the psychotic and nonpsychotic groups are shown in table III, and differ significantly (Fisher Exact Probability = 0.0077).

It is most noteworthy that, 6 of these 7 cases were rendered seizure-free by their surgery as judged by a follow-up of more than 5 yrs. The 7th patient (diagnosis: schizo-affective psychosis) had only occasional seizures, that is to say less than 4/year. Despite the excellent relief of the seizures indicating accurate lateralization of the epileptogenic lesion, the psychiatric picture did not substantially improve in any case. One patient was of particular interest because, although he had been noted to have 'some psychiatric abnormality' preoperatively, it was not until 3 years after his successful surgery that a full-blown paranoid schizophrenia-like psychosis occurred. In 2 other patients, both with serious preoperative psychopathology, the psychiatric features appeared to progress despite the surgical relief of their seizures.

Although the hypothesized specific relationship between the presence of a *left* temporal lobe epileptogenic lesion and the occurrence of a so-called 'schizophrenia-like' psychosis appeared to have been confirmed in this second study, an important concern remained. That is, in these studies the determination of lesion laterality rested upon a variety (neurological, psychiatric, radiological, neuropsychological and electrophysiological) of data. Despite the wealth of data such studies provide, there often remains some doubt about the certainty of the epileptogenic lesion's laterality [23]. Consequently, in a *third study,* we adopted a different strategy in order to reexamine our hypothesized association of psychosis with left TLEL. Instead of limiting our judgment regarding the laterality of the epileptogenic lesion to an analysis of the sort of data cited above, we proposed that the most rigorous criterion for correct lateralization was the successful *relief* of epilepsy following surgical excision. An analysis of the psychiatric diagnoses made in such a population seemed to provide the basis for making the most confident statement about the specificity of the association of psychosis and the laterality of the epileptogenic lesion. The data of this third study derive from a record review of the cases investigated at the Unité de recherches sur l'épilepsie (Paris), between 1959 and 1979.

As in the second study [25] the patients were classified with respect to the effect of the surgery on subsequent seizure frequency. At the time of analysis the mean postoperative duration was 5.6 years; and no patient with less than 1 year of follow-up was included in the study. Patients with no seizures following surgery were designated class 1; those with rare seizures (<4/year) as class 2; those with a significant reduction in seizure frequency (but >4/year) as class 3; and those with no significant improvement as class 4. For the purposes of this study, the epileptogenic lesions were considered to have been correctly lateralized only in those patients with class 1 or 2 outcomes. In reviewing the medical records, only those patients who had previously been given a *diagnosis* of psychosis, at some point in their illness, were accepted as cases of psychosis for the purposes of the current study.

The plan of this third study was first to determine the proportion of left and right temporal excisions, resulting in *relief* of seizures, that had been performed on a group of epileptics whose detailed workups indicated a unilateral epileptogenic lesion. Secondly, we attempted to map onto this left/right distribution the psychiatric diagnoses (specifically psychosis) that had been made and *recorded* in the patients' medical histories. Thus, this retrospective approach precluded our introduction of any bias with respect to either the selection of the surgical candidates or the choice of the operative procedures. That is to say, these decisions had already been made prior to the inception of this investigation. Similarly, to eliminate any bias regarding the psychiatric aspects, we made no attempt to authenticate the validity of those psychiatric diagnoses that had been recorded, i.e., we did not rediagnose these cases.

In the Unité series, of more than 600 patients studied, 101 cases were found who had been considered to have a temporal lobe epileptogenic lesion established with sufficient certitude to undergo a surgical excision. The composition of the entire series is shown in table IV. Of the 101 cases, 80 (group A) met the criteria for further analysis, i.e., the surgery resulted in the relief of the seizure disorder (classes 1 and 2) and the medical record was sufficiently detailed, to evaluate the psychiatric history of the patient.

In group A, left temporal excisions had been performed in 20 cases and right temporal excisions in the remaining 60. The preponderance of patients with right-sided excisions is highly significant (Binomial Test: $p < 0.0001$). This is of interest because it stands in sharp contrast to our first study [24] in which, as will be recalled, we sought patients who exhibited *coexistent* medically intractable epilepsy and severe psychopathology. In that study patients with left-sided lesions predominated. In the epilepsy program from which the data of this third study derive, however, a deliberate attempt had been made to exclude patients who demonstrated a serious behavior disorder, or *active* psychosis, at the time of intake. This was necessitated by the requirement for extreme cooperation during the long and tedious, detailed studies. In some patients the epileptogenic lesion extended far posteriorly, and we therefore considered the possibility that this left/right difference might also be accounted for by the surgeon's reluctance to operate in the latter cases, for fear of producing an aphasia.

Consequently we examined the interaction between left/right and anterior/posterior excisions (table V). The contingency coefficient was not significant ($cc = 0.103$), i.e., there was no statistically significantly, greater likelihood of *not* operating on left posterior lesions. To further pursue the probability of this explanation (i.e., fear of producing an aphasia), we recalculated it after excluding all the posterior cases. Even when only the cases with anterior excisions are considered (i.e., when aphasia was not a major concern) right-sided cases still significantly predominate (Binomial Test: $p < 0.004$). As in other reports concerning the surgical treatment of epilepsy [25, 31], males tended to predominate in this sample, but not significantly so (table VI). Similarly, when the coefficient of contingency is calculated for the laterality/gender interaction, the result, $cc = 0.131$, is also not statistically significant. It should be noted, however, that women

Table IV. Classification of the surgical patients (n = 101)

Group A	
Classes 1 and 2	80
Group B	
Classes 3 and 4	14
Lost to follow-up or dead	7

Table V. Group A: surgical excisions (n = 80)

	Right	Left
Posterior	25	6
Anterior	35	14

Contingency coefficient = 0.103 (NS)

Table VI. Gender vs. laterality group A (n = 80)

Females: males = 33:47   Z = 1.45 (NS)
Females: males/right: left   cc = 0.131 (NS)

Females   left:right = 6:27, $p < 0.0006$
Males   left:right = 14:33, $p < 0.01$

undergoing left temporal excisions constitute only a small proportion of the total. We therefore considered the sexes separately (lower half of table VI). In that case, the Binomial Test, assuming no a priori laterality bias, reveals that there is a small but significant ($p < 0.01$) left/right difference for men. The difference for women, however, is striking, $p < 0.0006$. These data also reinforce the contention that the observed left/right disproportion cannot be explained solely on the issue of aphasia, i.e., there is no reason to believe that this risk is greater in women. This finding, as will subsequently be detailed, also has relevance for the problem of psychosis per se.

The attempt to exclude patients with active psychopathology from this series notwithstanding, we found 7 cases with unilateral epileptogenic lesions warranting a temporal excision, who were diagnosed at some point

Table VII. Laterality vs. psychosis: group A (n = 80)

	Psychotic	Nonpsychotic
Right	2	58
Left	5	15

$0.01 > p > 0.005$

Table VIII.

	Dominance		Handedness	
	nondominant	dominant	right	left
Nonpsychotic	59	14	68	5
Psychotic	3	4	2	5
	cc = 0.20, NS		cc = 0.44, p < 0.001	

Table IX. Nonoperated group (n = 46)

	Total patients	Males	Females
Ratio left:right	37:9	17:6	20:3
p	<0.0001	<0.04	<0.001

in their histories as having been psychotic. It was not possible to ascertain with certainty the latency between the two disorders. However, in all the psychotic patients, the history of seizures antedated the psychiatric disorder. Of these 7 cases, 2 had undergone a right and 5 a left temporal excision. The relative frequencies of left- and right-sided epileptogenic lesions in the psychotic and nonpsychotic groups are shown in table VII, and differ significantly $(0.01 > p > 0.005)$. Following surgery the psychiatric picture was not substantially modified despite the excellent relief of the seizures indicating accurate lateralization of the epileptogenic lesion.

The fact that *left*-sided lesions predominated in the psychotic group, raised the obvious question of whether the specific association was between psychosis and epileptogenic lesions involving the left side per se, or rather the *dominant* hemisphere. Therefore, utilizing the results of the preoperative Wada test, we looked at the interaction between dominant/nondominant hemisphere excisions and psychosis/nonpsychosis. From table VIII, it may be seen that the contingency coefficient for this interaction was not significant. However, when we looked at handedness, rather than cerebral dominance *at the time of operation,* an entirely different picture emerged. The interaction between handedness and psychosis (table VIII) demonstrates that amongst the nonpsychotic patients, sinistrals account for <7% of this group, i.e., about what might be expected in a normal population. By contrast, for the psychotic patients, sinistrals account for >70% of the group, a highly significant difference ($p < 0.001$).

The data presented to this point, for reasons detailed above, derive only from those patients with medically intractable epilepsy who came to *surgery,* i.e., a very special subset of the population of such patients. We wondered, therefore, whether similar differences, i.e., with respect to laterality and gender, also occurred amongst those patients who did not come to surgery. To assess this we reviewed the entire Unité series once again in an attempt to find patients who, based on clinical, neuroradiological and scalp EEG data were thought to have unilateral temporal lobe epileptogenic lesions, but who did not undergo surgery.

46 such patients were found who, for a variety of reasons, did not come to surgery. These included: (a) patients declining the procedure when fully informed of the details of the surgery; (b) inadequate therapeutic (anticonvulsant) trials; (c) epilepsy of brief duration in very young patients; (d) intercurrent medical problems, and (e) severe psychopathology. The composition of this nonsurgical group with respect to the laterality of the epileptogenic lesions is shown in table IX.

The preponderance of left-sided lesions in this group stands in obvious contrast to the reciprocal finding in the surgical group. It should be noted, however, that this preponderance of left-sided lesions is largely due to its prevalence in the *female* patients, a point to which we shall return. In an attempt to ascertain the interaction of laterality and gender in the 'general' population of intractable epileptics from which these surgical and nonsurgical samples were drawn, we combined the two into a synthesized, theoretical 'parent population'. From table X it can be seen that the coefficient of contingency for this interaction does not reach a level of significance. We next looked at the effect the decision to operate had on this parent popu-

Table X. Synthesized theoretical 'parent-population' (n = 126)

	Females	Males
Right	30	39
Left	26	31

cc = 0.022 (NS)

lation with regard to these factors, i.e., laterality and gender. By comparing tables VI and X, it can be seen that for right-sided cases males and females are about equally likely to come to surgery, i.e., 85 vs. 90%. With respect to left-sided cases, this comparison indicates that females are likely to be operated upon only about half as frequently as males. We believe these findings also argue against the idea that left-sided cases are underrepresented in the surgical group solely as a consequence of the surgeons' fear of producing an aphasia in such cases. That is to say, there appears to be no reason why such a fear should be greater in women than in men.

It is fully acknowledged that the determination of the laterality of the epileptogenic lesions in the nonsurgical cases rests upon less rigorous criteria than is the case for the surgical patients. Therefore, any correlations derived between lesion laterality and clinical features in the former group must be interpreted with great caution. With this caveat clearly in mind, we were nevertheless struck by the observation that more than half (19/37) of the 'left-sided' nonsurgical cases had major psychiatric problems (including 2 cases diagnosed as schizophrenics). By contrast, less than one quarter of the 'right-sided' cases (2/9) had major psychiatric problems, and there were no instances of psychosis.

For the strategic reasons stated at the outset, this third study was based on data derived from patients who had surgery for a *temporal lobe* epileptogenic lesion. Much controversy exists as to whether patients with 'temporal epilepsy' are more likely to develop psychiatric problems as compared to those with other forms of epilepsy. Since it is virtually impossible, with perhaps the exception of true petit-mal, to rule out a focal origin in cases of so-called 'primary generalized epilepsy', we compared our results with data derived from patients at the Unité who had had excisions for other focal, *nontemporal,* epileptogenic lesions. In doing so it is essential to point out that in speaking of temporal lobe epileptogenic *lesions,* we are not speaking of 'temporal lobe *epilepsy*', 'temporal lobe *seizures*', 'psychomotor *seizures*', 'complex partial *seizures*', or any other description of the *clinical features* of these patients' seizure disorder! We found 42 patients who met our criteria with respect to surgical outcome and completeness of their psychiatric histories. Of these 42 cases, 13 had frontal excisions, 10 had parietal excisions, 11 had occipital excisions and 8 were classified as central excisions. Of these patients, *none* had a history of psychosis. We found one other patient (unoperated) who, based on the presurgical studies, was thought to have a left occipital epileptogenic lesion, and who had been diagnosed as schizophrenic. Since this patient did not come to operation she is not included in the series.

In agreement with other recent studies [19], the results of our investigations indicate that patients with left-sided temporal lobe epileptogenic lesions are at special risk to develop a schizophrenia-like psychosis. The finding that sinistrals constituted the majority in the psychotic group is congruent, too, with similar observations in still other studies [11, 13]. Taken together with our data from the preoperative Wada tests, we interpret the latter finding to indicate that those patients developed a left-sided epileptogenic lesion early (perhaps perinatally) in life. A consequence of this is a shift from right to left handedness and in some patients results in a shift to right cerebral dominance.

Discussion and Critique

All of the data cited above, including our own, derive from retrospective studies of patients seen at centers devoted to the surgical treatment of epilepsy. It might be argued that these findings are peculiar to this special subset of the population of temporal lobe epileptics. However, against this are the strikingly similar observations of *Lindsay* et al. [15] in their truly remarkable prospective, nonsurgical study. They found that in a cohort of 87 children with temporal lobe epilepsy followed (for over 15 years) into adulthood, 9 patients had developed 'the schizophreniform psychosis of temporal lobe epilepsy'. Moreover, they noted that 'none of the psychotics had a right-sided focus, 7 had a left-sided focus and 2 had bilateral discharges'.

An important finding in our second and third studies was the observation that women with left-sided epileptogenic lesions tended to be relatively underrepresented in the temporal surgical group. Moreover, in our third study they were found to be overrepresented in the unoperated group. We believe this is noteworthy for two reasons. First, as pointed out above, it indicates that the fear of producing an aphasia is an insufficient explanation to account for the totality of the observed left/right disproportions of the surgical and nonsurgical temporal groups. Secondly, the relative underrepresentation of women with left-temporal epileptogenic lesions in the surgical group is of interest because *Taylor* [30], as noted above, has postulated that this is the group most at risk for developing the schizophrenia-like psychosis associated with epilepsy. Based on that proposition, the observed underrepresentation of women with left-sided lesions is precisely what would have been anticipated in the situation where, as was the case in both the second and third studies, psychotic patients were deliberately excluded from a sample (surgical-temporal group) of epileptics.

In addition to our interest in the effect of the laterality of the epileptogenic lesion on the prevalence of psychosis, we were curious to know whether patients with *temporal lobe* epileptogenic lesions are, in general, at greater risk for this complication than are patients with other focal forms of epilepsy. The data of the third study suggest that psychosis is a rare complication in patients with other focal (nontemporal) epileptogenic lesions, and thus it appears to be relatively specific for patients with temporal lobe epileptogenic lesions. *Stevens* [28], too, has commented on its rarity in patients with other forms of focal epilepsy. The specificity of this relationship was first emphasized by *Gibbs* [4]. In a study of 275 cases with focal epilepsy, he found that the prevalence of psychosis amongst patients with temporal lobe epileptogenic (EEG) foci was 17%. By contrast, psychosis occurred in only 2% of patients with 'occipital' foci and did not occur at all in patients with other focal epileptogenic lesions. These relative frequencies are strikingly similar to the findings in our third study.

The majority of studies [for review see *Herman,* 9], which conclude that there is no difference in the propensity for patients with temporal lobe epilepsy and those with other forms of epilepsy to develop psychiatric complications, have been those which compared patients with psychomotor seizures on the one hand and those with generalized seizures on the other. It is well known that patients with focal *epilepsy* may often have generalized *seizures.* By contrast patients with 'primary generalized' *epilepsy* do not have focal *seizures.* The absence of focal seizures is a necessary but insufficient criterion for making the diagnosis of 'primary generalized' epilepsy. Consequently, errors in diagnosis, which are bound to occur, can only occur in one direction. That is to say, some patients with TLEL who may have only generalized *seizures* will be mistakenly diagnosed as having generalized *epilepsy*. The obverse, i.e., making the diagnosis of focal (temporal lobe) epilepsy in patients who have 'primary generalized' epilepsy, will not occur. This inevitability for systematic error must, in part at least, account for studies reporting no differences in the psychiatric features in patients with temporal versus generalized epilepsy.

It should be obvious that the actual overall prevalence of psychosis in our current surgical series, may have been over- or underestimated as a result of our decision to accept the previously recorded psychiatric diagnoses. However, it must be equally obvious that there is absolutely no reason to believe that the accuracy of those diagnoses would have systematically varied as a function of which temporal lobe was ultimately excised. This is not a trivial point, and restated: it is not that the absolute prevalence of psychosis in patients with temporal lobe epileptogenic lesions is not an important issue, but rather that it is irrelevant to the fundamental question of our studies, i.e., the proportion of the psychoses associated with either left- or right-sided epileptogenic lesions. However, the prevalence of 11% observed in our second study is in remarkably close agreement with the 9% found in our third study, and the 10% prevalence reported by *Lindsay* et al. [15] in their prospective study – see also *Guerrant* et al. [8], 13%; *Pritchard* et al. [22], 11%; *Falconer* [2], 14%. It is suggested, therefore, that our findings may well be valid estimates of the actual prevalence, and that *Gibbs'* [4] 30-year-old assessment of 17%, although perhaps slightly high, has in fact stood the test of time.

Why patients with epileptogenic lesions of the left temporal lobe should be at special risk for developing a psychosis is not immediately apparent. If, however, our postulate is correct that it is epilepsy commencing in the *dominant* temporal lobe that is the critical factor, then it may well be that thought processes, which are so intimately involved with language and symbolic representation, are disrupted by the epileptic process and thus result in a thought disorder. It has been speculated too that that group of disorders called 'The Schizophrenias' and temporal lobe epilepsy may share disordered function of certain dopaminergic substrates [27]. In this regard it is of interest that, in studying a group of *left* temporal lobe epileptics, *Peters* [21] recently reported finding a significant difference in CSF levels of the dopamine metabolite, homovanillic acid, in those patients with and those without an associated schizophrenia-like psychosis.

In our studies, the psychosis associated with epilepsy has deliberately been referred to as 'schizophrenia-like'. This is so because despite certain differences it is on balance phenomenologically more like than different from schizophrenia. In a recent study, *Perez and Trimble* [20] reported that 11 out of 12 temporal lobe epileptics classified as schizophrenic met the criteria for the diagnosis of nuclear schizophrenia, based on Schneider's first rank symptoms. They concluded that 'from a diagnostic point of view epileptics and nonepileptics classified as schizophrenic are psychopathologically comparable'. The criteria for making the diagnosis of 'schizophrenia-like psychosis associated with epilepsy' is considered in more detail in the chapter by *Trimble and Perez* [this volume].

Obviously, even if the two disorders should prove to share a certain common pathophysiology, there is no a priori reason why they must be clinically identical. In their classic report, *Slater* et al. [26] noted that a clinical feature distinguishing this disorder from 'true schizophrenia' appeared to be the preservation of affect. Another feature which appeared to be of distinguishing value, in our cases, was the less socially disruptive character of the psychosis. That is to say, the psychosis is often more periodic, usually does not require the use of major tranquilizers, and most often does not require chronic hospitalization.

These studies raise two other issues which deserve comment. *Falconer* [2] reported that in his series, the schizophrenia-like psychosis appeared to be associated with hamartomatous lesions of the temporal lobes. By contrast, he found aggressivity tended to be associated with the more commonly found lesion, i.e., mesial temporal sclerosis. The specificity of these associations were not confirmed in the study by *Jensen and Klinken* [10] and is not supported by our own, albeit, limited histopathological data. It is noteworthy that patients with mesial temporal sclerosis tend to have the onset of their epilepsy in early childhood; and patients with hamartomas, on the other hand, generally do not begin to have seizures until puberty. Consequently, it may be as *Taylor* [29] has suggested that it is the age of onset of the epilepsy rather than the histology of the underlying lesion which determines the specific type of psychopathology that may supervene.

Our own data are insufficient to allow us to choose between these points of view. However, it may not be totally unreasonable to speculate that since a major aspect of early childhood socialization is the development of inhibitory mechanisms controlling limbically mediated behavior, early onset epilepsy associated with mesial temporal sclerosis may interfere with this process. As a result of this, there may develop only poor control over aggressive tendencies. In fact, aggressive tendencies might actually be reinforced by abnormal electrical discharges arising in the limbic system. It is suggested, too, that the more 'sophisticated' mental mechanisms, viz., those subsuming paranoid ideation (a common feature of the schizophrenia-like psychosis) probably are not operative in infancy but may be mobilized in late-onset epilepsy associated with hamartomas. The accuracy of these speculations notwithstanding, the importance of determining the relative significance of these possible risk factors (i.e., lesion type, age-at-onset, and duration of the epilepsy) with regard to the selection criteria for and timing of possible neurosurgical intervention is obvious.

A last point requiring comment is that we [25] and others [12] have observed that the onset of psychosis might *follow* the successful surgical elimination of the epilepsy. One question raised by these observations deals with the relative importance of the contributions made by the seizures on the one hand, and the underlying anatomical lesions on the other, to the psychiatric problems. In keeping with *von Meduna's* [17] theory of a mutual antagonism between seizures and schizophrenia, it may be that seizures are, in some way, protective against the emergence of a psychosis. If this is so, it is not immediately obvious why the 'postoperative' psychosis may not appear until some *years* after the elimination of the seizures. Moreover, such a mechanism by itself would not account for the majority of cases in which the psychosis appears in the midst of the seizure disorder. It is, of course, possible that the postoperative psychoses would have developed, in time, had no surgery been performed in such cases. It is perhaps equally possible that, following long-standing epilepsy, an additional brain insult (temporal lobectomy) might actually facilitate the emergence of a psychosis.

Fundamentally, this dilemma reflects our incomplete understanding of the natural history of this disorder. In part, this derives from the practice of using patients with bilateral foci, who thereby do not meet the strict criteria for surgery, as controls for surgical candidates viz.: those with unilateral foci. Given the obvious ethical issues involved, it is most likely that pooled data from sizable, carefully controlled studies of such cases will be required if we are to attempt to resolve these important questions.

References

1 Epilepsy Foundation of America: Basic statistics on the epilepsies, pp. 59–61 (Davis, Philadelphia 1975).
2 Falconer, M.A.: Temporal lobe resection for epilepsy and behavioral abnormalities. New Engl. J. Med. *289:* 451–455 (1975).
3 Flor-Henry, P.: Schizophrenic like reactions and affective psychoses associated with temporal lobe epilepsy: etiological factors. Am.J. Psychiat. *126:* 148–152 (1969).
4 Gibbs, F.A.: Ictal and non-ictal psychiatric disorders in temporal lobe epilepsy. J. nerv. ment. Dis. *11:* 522-528 (1951).
5 Gloor, P.; Oliver, A.; Ives, J.: Prolonged seizure monitoring with stereotaxically implanted depth electrodes in patients with bilateral interictal temporal epileptogenic foci: how bilateral is bitemporal epilepsy? (Abstract). 10th Epilepsy International Symposium, Vancouver, p. 22 (1978).
6 Gowers, W.R.: The borderland of epilepsy (Churchill, London 1907).
7 Gowers, W.R.: Epilepsy and other chronic convulsive diseases. Reprint series (Dover, New York 1964).
8 Guerrant, J.; Anderson, W.W.; Fischer, A.; Weinstein, M.R.; Jaros, R.M.; Deskins, A.: Personality in epilepsy, pp. 27–99 (Thomas, Springfield 1962).
9 Herman, B.P.; Deficits in neuropsychological functioning and psychopathology in persons with epilepsy: a rejected hypothesis revisited. Epilepsia *22:* 161–167 (1981).
10 Jensen, I.; Klinken, L.: Temporal lobe epilepsy and neuropathology. Histological findings in resected temporal lobes correlated to surgical results and clinical aspects. Acta neur. scand. *54:* 391–414 (1976).
11 Jensen, I.; Larsen, J.K.: Mental aspects in drug resistant temporal lobe epilepsy (Abstract). 10th Epilepsy International Symposium, Vancouver, p. 527 (1978).
12 Jensen, I.; Larsen, J.K.: Mental aspects of temporal lobe epilepsy, follow up of 74 patients after resection of a temporal lobe. J. Neurol. Neurosurg. Psychiat. *42:* 256–265 (1979).
13 Kristensen, O.; Sindrup, E.H.: Psychomotor epilepsy and psychosis. I. Physical aspects. Acta neur. scand. *57:* 361–369 (1978).
14 Lennox, W.G.: in Hunt, Personality and the behavior disorders, vol. 2 (Ronald Press, New York 1944).
15 Lindsay, J.; Ounsted, C.; Richards, P.: Long term outcome in children with temporal lobe seizures. III. Psychiatric aspects in childhood and adult life. Devl. med. Child Neur. *21:* 630–636 (1979).
16 Masland, R.L.: Commission for the control of epilepsy (guest editorial). Neurology, Minneap. *28:* 861–863 (1978).
17 Meduna, L. von: General discussion of the cardiazol therapy. Am. J. Psychiat., suppl. 94, pp. 40–50 (1938).
18 Parsonage, M.: Discussion of inter-ictal manifestations of complex partial seizures; in Penry, Daly, Advances in neurology, vol. 2, p. 111 (Raven Press, New York 1977).
19 Perez, M.M.; Trimble, M.R.: Different kinds of psychosis as related to different kinds of epilepsy (Abstract). 12th Epilepsy International Symposium, Copenhagen (1980).
20 Perez, M.M.; Trimble, M.R.: Epileptic psychosis – diagnostic comparison with process schizophrenia. Br. J. Psychiat. *137:* 240–245 (1980).
21 Peters, J.G.: Dopamine, noradrenaline and serotonin spinal fluid metabolites in

temporal lobe epileptic patients with schizophrenic symptomology. Eur. Neurol. *18:* 15–18 (1979).
22 Pritchard, P.B.; Lombroso, T.L.; McIntyre, M.: Psychological complications of temporal lobe epilepsy. Neurology *30:* 227–232 (1980).
23 Rossi, G.F.: Closing remarks: surgery of epilepsy; in Gillingham, Gybels, Hitchcock, Rossi, Szikla, Advances in stereotactic and functional neurosurgery. Acta neurochir. suppl. 30, pp. 25–34 (Springer, New York 1980).
24 Sherwin, I.: Clinical and EEG aspects of temporal lobe epilepsy with behavior disorder, the role of cerebral dominance. McLean Hosp. J. (special issue), June, pp. 40–50 (1977).
25 Sherwin, I.: Psychosis associated with epilepsy: significance of the laterality of the epileptogenic lesion. J. Neurol. Neurosurg. Psychiat. *44:* 83–85 (1981).
26 Slater, E.; Beard, A.W.; Glithero, E.: The schizophrenia like psychoses of epilepsy (i-v). Br. J. Psychiat. *95:* 109–150 (1963).
27 Stevens, J.R.: An anatomy of schizophrenia? Archs gen. Psychiat. *29:* 177–189 (1972).
28 Stevens, J.R.: Interictal clinical manifestations of complex partial seizures. Adv. Neurol. *11:* 85–112 (1975).
29 Taylor, D.C.: Aggression and epilepsy. J. Psychosom. Res. *13:* 229–236 (1969).
30 Taylor, D.C.: Factors influencing the occurrence of schizophrenia-like psychosis in patients with temporal lobe epilepsy. Psychol. Med. *5:* 249–254 (1975).
31 Taylor, D.C.; Falconer, M.A.: Clinical, socio-economic and psychological changes after temporal lobectomy for epilepsy. Br. J. Psychiat. *114:* 1247–1261 (1968).
32 Temkin, O.: The falling sickness; 2nd ed., pp. 265–270 (Johns Hopkins Press, Baltimore 1971).

I. Sherwin, MD, Department of Neurology, Harvard Medical School,
V.A. Hospital (151), Bedford, MA 01730 (USA)

The Phenomenology of the Chronic Psychoses of Epilepsy

Michael R. Trimble[a], Margarita M. Perez[b,1]

[a]Consultant Physician in Psychological Medicine, The National Hospitals for Nervous Diseases, London, England; [b]Senor Registrar, St. Georges Hospital, London, England

Introduction

In an earlier chapter, the history of the relationship between epilepsy and psychosis was presented, and some early descriptions of the symptomatology of these states reviewed. In view of the growing recognition and importance of the chronic psychotic disorders that occur in epilepsy, and the growing speculation that their aetiology is related to disturbances of temporal lobe/limbic system structures, it seems of utmost importance to describe as accurately as possible the phenomenology of these patients. As has been seen, all of the papers referred to used clinical impressions to define patients' symptoms and no attempt was made to systematically evaluate phenomenology.

Measurement of Psychiatric Symptoms

In psychiatry over the past 20 years more definitive methods have been adopted for quantification of psychopathology, which mainly use rating scales in an attempt to increase reliability of clinical observation. These have basically been of two sorts, either self-assessment questionnaires filled in by patients, or interviewer-rated scales which provide working definitions of psychopathology in order to compare patterns of symptoms in different populations. A more recent development in this field has been the

---

[1] We gratefully acknowledge the help of the Brain Research Trust, who supported our studies discussed in this article.

use of structured or standardised interviews, in which the manner in which the symptoms are elicited, as well as the way in which they are recorded, is laid down. Questions are asked in a specific order, and ratings of patients' answers are made based on standardised definitions. Two of the most widely used of these are Spitzer's Psychiatric Status Schedule [18] and the Wing Present State Examination (PSE) [12]. The latter is used more commonly in Great Britain, and has, over the years, undergone a number of developments and refinements. The interview technique is based on the examination of the patient, and the clinician decides whether a particular symptom is present or not after all the questions laid down, and appropriate probes, have been asked. The questions relate to the patient's experiences during the month prior to the interview, and this of necessity does exclude most of the psychiatric history, which becomes the subject of a separate enquiry. The 9th edition consists of a check-list of 140 items, which has been condensed from over 400 in earlier versions of the test. It is important to note that the interview is still basically clinical, the schedule not being like a questionnaire. It has reliability and has been shown to give good agreement, particularly for the diagnosis of schizophrenia, in many countries of the world, including America, Taiwan, Czechoslovakia and Nigeria [11]. It is of particular value for the classification of nuclear schizophrenia, a syndrome based on Schneider's first-rank symptoms.

The results from the PSE schedule are analysed by a computer program known as the Catego, in which the original items pass through a progressive series of condensations, and decisions about the actual diagnosis are postponed until the final stage. The item content is thus reduced to a number of 'symptoms', which are then further refined to 38 basic 'syndromes'. These range from those measuring general and non-specific anxiety, to neurosis, depression, delusions and hallucinations, and include measures of general attitude and behaviour, observable mood disturbances, and motor symptoms. Mean syndrome profiles can be derived from the PSEs to compare groups, or to monitor change with time. The Catego program prints out the syndromes with a 3-point ranking, indicating the degree of certainty with which it may be said to be present.

The final output of the program is an allocation to 1 of 50 subclasses, which are collapsed into broader classes. Descriptions of these are given in *Wing* et al. [12]. In general, each PSE syndrome profile is allotted to one class, which is basically descriptive, and represents a summary of the PSE ratings [13].

Although this technique has found its main use in identifying and classifying psychiatric symptomatology, particularly in comparison of international differences and similarities in diagnosis, it seemed the appropriate technique for trying to clarify in detail the phenomenology of psychoses associated with epilepsy, and has been used by us for this purpose [6].

PSE Evaluation of the Chronic Psychoses of Epilepsy

24 patients, consecutively referred to the National Hospitals suffering from unequivocal epilepsy and active psychosis which had continued for more than a month and in whom consciousness was clear, were examined using the PSE. The results were compared with a group of 11 schizophrenic patients who were previously assessed by two psychiatrists, both of whom agreed the diagnosis was schizophrenia. In the analysis of the results Fisher's Exact Probability Test was used. Of the 24 epileptic patients, 12 received a class diagnosis of schizophrenic psychosis, and of these analysis of the subclass revealed 11 who were classified as nuclear schizophrenia (table I); 10 were classified as having some kind of affective disturbance, and 1 as a paranoid psychosis. In contrast to this group, 9 of the patients with schizophrenia were classified as nuclear schizophrenia, 1 as paranoid psychosis, and 1 as psychotic depression.

In a further analysis, the type of psychosis present in those patients classified clinically as having complex-partial seizures of probable temporal lobe origin was compared with those having generalised epilepsy. As can be seen from table II, a classification of nuclear schizophrenia was exclusively associated with the temporal lobe group, the other diagnostic categories being present in both groups. 1 patient who had a diagnosis of schizophrenia without first-rank symptoms came from the generalised group. A more comprehensive account of these data is given in *Perez and Trimble* [6].

Figure 1 shows the PSE syndrome profiles of those patients classified as having schizophrenia on the PSE comparing the patients with epilepsy to those without. In this figure the syndromes have been grouped together in the various categories as shown on the horizontal axis, the vertical axis relating to the percentage of patients who demonstrate the particular syndrome. It can be seen that the profiles of the two groups are very similar, the only significant differences relating to visual hallucinations, which were not found in the patients with epilepsy, and grandiose delusions which included religious delusions that were present significantly more in

Table I. Catego subclasses and classes – diagnostic comparison between epileptic psychotic and schizophrenic patients

Subclasses diagnoses	Epileptics (n = 24)	Schizophrenics (n = 11)	Classes diagnoses	Epileptics (n = 24)	Schizophrenics (n = 11)
Nuclear schizophrenia	11	9	Schizophrenic psychosis	12	9
Schizophrenia without first rank symptoms	1	–	Manic psychosis	3	–
Residual schizophrenia/mania	1	–	Depressive psychosis	3	1
Manic psychosis	3	–	Paranoid psychosis/retarded depression	3	–
Psychotic depression	3	1	Paranoid psychosis	2	1
Paranoid psychosis/retarded depression	3	–	Borderline psychosis	1	–
Paranoid psychosis/affective psychosis	1	–			
Paranoid psychosis	1	1			
Total	24	11	Total	24	11

Table II. Catego subclasses and classes – diagnostic comparison between psychotic patients with temporal lobe (TLE) and with generalised (GEN) epilepsy

Catego subclasses	TLE (n = 17)	GEN (n = 7)	Catego classes	TLE (n = 17)	GEN (n = 7)
Nuclear schizophrenia	11	–	Schizophrenic psychosis	11	1
Schizophrenia without first rank symptoms	–	1	Manic psychosis	1	2
Residual schizophrenia/mania	–	1	Depressive psychosis	1	2
Manic psychosis	1	2	Paranoid psychosis/retarded depression	3	–
Psychotic depression	1	2	Paranoid psychosis	1	1
Paranoid psychosis/retarded depression	3	–	Borderline psychosis	–	1
Paranoid psychosis/affective psychosis	1	–			
Paranoid psychosis	1	1			

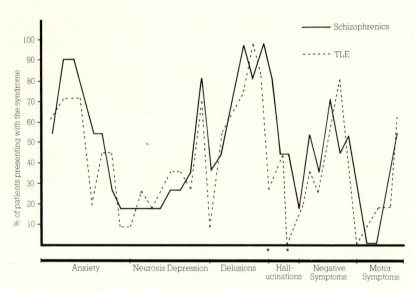

Fig. 1. PSE syndromes. Profile in schizophrenic and psychotic TLE patients. *p < 0.05 for grandiose delusions and visual hallucinations.

the process schizophrenic patients. It should be noted that simple depressive syndromes, which include inefficient thinking, depressed mood, hopelessness, suicidal ideas and depression were prominent though not universal in both groups. In addition, syndromes measuring general anxiety, motor disturbances, general behaviour and attitude during interview, were also similar, and auditory hallucinations of the non-depressive kind were present in the same proportion in both groups.

In contrast, patients with generalised epilepsy do not show nuclear schizophrenic symptoms, and the psychoses in these cases were mainly confined to affective, paranoid, or borderline categories. Syndrome profile examination of these patients were characterised by general anxiety and mood disturbances. Depression, in particular resembling the endogenous type, was the commonest, although mood elation and fantastic delusions were also noted. Visual hallucinations were absent and the only auditory hallucinations recorded were those of a depressive kind. Delusions were of reference and of persecution, and neurotic symptoms were noted to be more prominent in these patients than in the complex-partial seizure group.

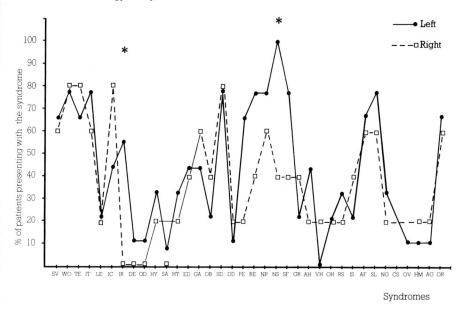

Fig. 2. Comparison of syndrome profiles between left (n = 9) and right (n = 5) sided foci in psychotic temporal lobe epileptic patients. *p < 0.02, nuclear syndrome; *p < 0.06, ideas of reference. For explanation of syndromes, see *Wing* et al. [12].

There has been a growing interest, particularly since the paper of *Flor-Henry* [2], in the effects of laterality on the clinical presentation of the psychoses and *Sherwin* [this volume] has discussed several aspects of this. Figure 2 shows the syndrome profiles of those patients with left-sided or predominantly left-sided lesions, compared with those who have right-sided lesions. In this figure the various syndromes are referred to using the standardised notation of the Catego system [12]. Two significant differences of importance should be noted. Patients with the left-sided lesions show significantly more ideas of reference and nuclear syndrome than the right-sided patients. This latter result may be all the more significant in that the patients who did rate as nuclear syndrome who had right-sided lesions were left-handed, and their dominance cannot therefore be accurately stated. Although not significant, it is important to note the increased occurrence of delusions of persecution, delusions of reference, fantastic delusions, and auditory hallucinations in the left temporal lobe patients and their absence of visual hallucinations.

Conclusions

These data lend support to the suggestions that the psychotic states associated with complex-partial seizures of temporal lobe origin vary in relationship to the laterality of the focus, and they must be seen in the light of other results which have attempted to clarify this issue. The data on laterality is reviewed in more detail in the chapter by *Sherwin,* [this volume], but it should be noted that a left-sided emphasis and a schizophreniform presentation has now been reported on by *Flor-Henry* [2], *Taylor* [9], *Toone and Driver* [10], *Sherwin* [7], and most recently by *Ounsted and Lindsay* [5]. There are, in addition, several studies that have come up with negative results in this regard, in particular those of *Bruens* [1], *Kristensen and Sindrup* [4] and *Jensen and Larsen* [3]. However, none of these latter authors examined in detail the phenomenology of their cases as noted, using clinical criteria for assessment. The authors that have demonstrated a laterality effect have tended to be British, and their diagnostic criteria for the assessment of the schizophreniform psychosis has been based on Schneiderian principles. Our own results lend further weight to the suggestion that a schizophrenia-like psychosis with Schneiderian first-rank symptoms occurs in epilepsy in relation to temporal lobe abnormalities more commonly than in other forms of epilepsy, and that when it does occur the likelihood is that the patient will have a dominant-sided lesion. It is our contention that conflicting data in this field have arisen because of the inexact descriptions of psychopathology that authors have attributed to patients, and it is suggested that future work should use techniques such as standardised and validated rating scales or semi-structured interviews for the gathering of clinical data. Without such methodology confusion is likely to persist, but with it the way forward may be clearer for a more precise linking of the psychopathological states observed in patients with epilepsy to the underlying neurological phenomena which we believe have relevance not only for the manifestation of the epilepsy, but also for the psychiatric illness.

References

1 Bruens, J.H.: Psychosis in epilepsy. Psychiatrica Neur. Neurochir. *74:* 175–192 (1971).
2 Flor-Henry, P.: Psychosis and temporal lobe epilepsy. Epilepsia *10:* 363–395 (1969).
3 Jensen, I.; Larsen, J.K.: Mental aspects of temporal lobe epilepsy. J. Neurol. Neurosurg. Psychiat. *42: 256–265 (1979).*

4 Kristensen, O.; Sindrup, E.H.: Psychomotor epilepsy and psychosis. Acta neurol. scand. *57:* 361–370 (1978).
5 Ounsted, C.; Lindsay, J.: The long-term outcome of temporal lobe epilepsy; in Reynolds, Trimble, Psychiatry and epilepsy, pp. 185–215 (Churchill-Livingstone, Edinburgh 1981).
6 Perez, M.M.; Trimble, M.R.: Epileptic psychosis. A psychopathological comparison with process schizophrenia. Br. J. Psychiat. *137:* 245–249 (1980).
7 Sherwin, I.: Psychosis associated with epilepsy. Significance of the epileptogenic lesion. J. Neurol. Neurosurg. Psychiat. *44:* 83–85 (1981).
8 Spitzer, R.L.; Endicott, J.; Robins, E.: Research diagnostic criteria. Archs gen. Psychiat. *35:* 773–782 (1978).
9 Taylor, D.: Factors influencing the occurrence of schizophrenia-like psychoses in patients with temporal lobe epilepsy. Psychol. Med. *5:* 249–254 (1975).
10 Toone, B.K.; Driver, M.V.: Psychoses and epilepsy; in Parsonage, Aspects of epilepsy (MCS Consultants 1980).
11 WHO: The international pilot study of schizophrenia (World Health Organisation, Geneva 1973).
12 Wing, J.K.; Cooper, J.E.; Sartorius, N.: The description and classification of psychiatric symptoms (Cambridge University Press, London 1974).
13 Wing, J.K.; Sturt, E.: The PSE-ID-CATEGO system. A supplementary manual (Institute of Psychiatry, London 1978).

MR. Trimble, MRCP, FRCPsych., Consultant Physician in Psychological Medicine, The National Hospital for Nervous Diseases, Queen Square, London WC1 3BG and M.M. Perez, MRCPsych. Senor Registrar, St. Georges Hospital, London (England)

Psychosis and Epilepsy: Similarities and Differences in the Anatomic-Physiologic Substrate

Robert G. Heath

Tulane University School of Medicine, New Orleans, La., USA

Introduction

For more than 30 years, the experimental and clinical research program of the Tulane University Department of Psychiatry and Neurology has focused primarily on the development of treatment for patients with psychiatric and neurologic disorders previously resistant to commonly used therapy. In the course of these physiologic and anatomic studies, new techniques evolved to permit us to explore brain activity and behavior simultaneously.

Particularly valuable data were obtained with use of deep and surface brain recordings in a series of 76 patients, all intractably ill, in whom the procedures were carried out for diagnostic and therapeutic purposes. Our techniques permitted the brain electrodes to remain accurately in position for 6 months to 2 years, long after artifacts resulting from anesthesia and implantation had subsided. Since the patients, fully conscious during the studies, were capable of reporting thoughts and feelings (unlike animals), we were able to make long-term observations of brain recordings while simultaneously monitoring mental activity and objectively observing behavior, thereby correlating recording changes with clinical phenomena. In the epileptic patients in the series, recordings were obtained during a variety of seizures.

Patient Population

All 76 patients in whom deep and surface electrodes were used were intractably ill, having failed to respond to other treatments. The 9 patients in the series who had epilepsy displayed behavioral disturbances ranging from inappropriate irritability to overt psychosis, in addition to having seizures. Two of the epileptic patients had only grand mal seizures, 4 had complex partial seizures along with grand mal, and 3 had complex partial seizures alone. Of the 63 patients with psychosis, 38 were diagnosed as schizophrenic and 25 had psychosis of other origin.

Methods

Techniques have been described for implantation of electrodes into precise brain sites, with the electrodes fixed at the skull so as to remain in place for prolonged periods [8, 9]. Our recording techniques, with use of Grass electroencephalographs have also been described in detail [8, 10]. Two or three machines were often synchronized to provide more channels of recordings, and in many instances, recordings were taped for later computer analysis. The methods we used for electrical and chemical stimulation of patients have previously been detailed [5, 6, 8].

Results

Epilepsy

Seizures. Correlating with the development of seizures in epileptic patients were certain consistent recording changes. Further, in individual patients, the nature of the recording changes was consistent from one seizure to another. In all the epileptic patients, distinct abnormalities were recorded at subcortical sites interictally, even when surface recordings from both cortex and scalp were within normal range. The abnormalities, in the form of intermittent spike and slow-wave activity, invariably occurred in the hippocampus and amygdala. When the patients were asymptomatic, the abnormalities were infrequent. As a seizure became imminent, the spike and slow-wave activity in the hippocampus and amygdala inevitably increased concomitantly with the patient's increasing restlessness and irritability.

In the minutes before onset of the aura preceding the clinical seizure, bursts of characteristic seizural activity occurred in these same deep structures. Then, within 30 s to 2 min before onset of the clinical seizure, the electrical seizure in the hippocampus and amygdala began on one side, and within seconds spread to the opposite side. In association with the aura, intermittent cortical spiking appeared, usually in the temporal cortex. Then, the seizural activity from the hippocampus and amygdala invariably intensified and spread to the septal region. The abnormalities were confined to these deep structures for 30s to 1 min, after which they spread

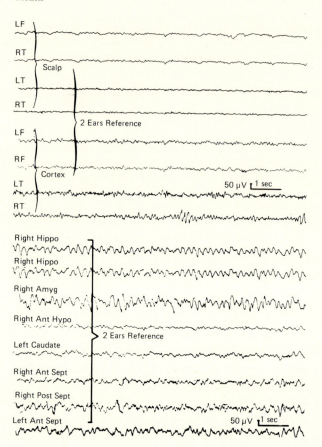

Fig. 1. Deep and surface recordings from an epileptic patient who suffered from seizures and episodic psychotic behavior when he was symptom-free.

to involve the entire brain concomitant with onset of the clinical seizure. Cessation of the clinical seizure was associated with a consistent recording pattern. Seizural activity stopped first in the hippocampus and amygdala and 3–5s later, in the septal region, and another 3–5s later, at the cortical sites. Generalized postictal slow activity then followed, gradually diminishing over a period of minutes. This activity was inevitably most pronounced and prolonged in the recordings from the septal region, subsiding gradually in correlation with the patient's return to full awareness (a 16-mm movie film, with voice narration, showing the recording changes during a seizure was made in the laboratory) (fig. 1, 2).

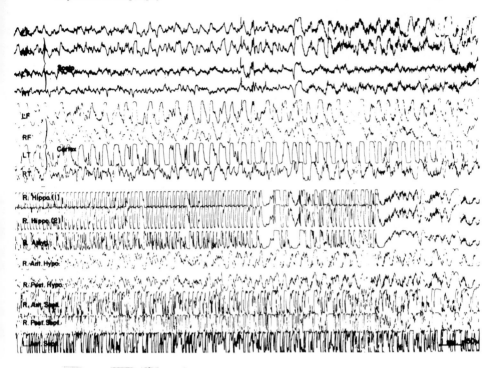

Fig. 2. Deep and surface recordings from the same epileptic patient during and just before termination of a seizure. Note that seizural activity stops first in the hippocampus and amygdala.

Disturbed Behavior. Severely disturbed behavior, including episodes of violence, are reputedly *not* characteristic of epilepsy. In our series of patients, however, some epileptics had such episodes, and four developed periods of intense rage in association with psychotic behavior. During rageful emotional dyscontrol, distinct recording changes, appearing principally in the hippocampus and amygdala, were characteristically high-amplitude spike and slow-wave activity interspersed with bursts of spindles at frequencies of 12–18 per s (fig. 3). As the emotional dyscontrol subsided, recordings returned to baseline. These recordings from epileptic patients shared some common characteristics with recordings of *non*epileptic patients during emotional dyscontrol. In the *non*epileptic patients, recording changes occurred at the same anatomic sites (hippocampus and

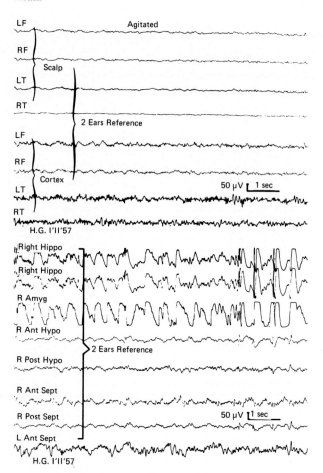

Fig. 3. Deep and surface recordings from the same epileptic patient when he was disturbed. The high-amplitude spike and slow wave activity was confined principally to the hippocampus and amygdala during episodes of dyscontrol.

amygdala). Spike and slow-wave activity was absent, however, with fast spindles occurring only during episodes of dyscontrol (fig. 4).

When epileptic patients developed psychotic behavior, recording changes closely resembled those obtained in psychotic patients *without* epilepsy. The characteristic recording change was a spread of the intermittent spike and slow-wave pattern from the hippocampus and amygdala to the septal region (fig. 5). Regardless of the underlying basic illness,

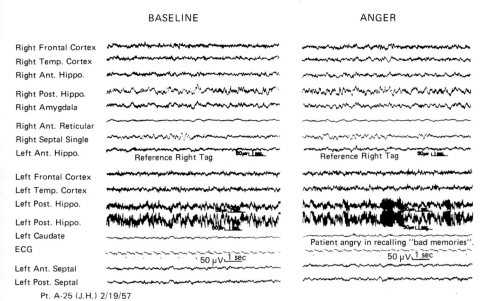

Fig. 4. Deep and surface recordings obtained from a nonepileptic, nonpsychotic patient. Note the intermittent fast spindles localized to the hippocampus when the patient expressed intense anger during a psychiatric interview.

psychotic behavior was correlated with septal spiking. In psychotic epileptic patients, slow-wave activity was more pronounced and spiking was more frequent and of higher amplitude than in the psychotic schizophrenic patients. In the epileptic patients, the onset of psychotic behavior was sometimes followed by seizures. In those instances, the chronology of recording changes was as described for seizures. When psychotic behavior remitted without a seizure, recordings returned to baseline.

Schizophrenia

Since our initial report in 1952 of spiking in recordings from the septal region of psychotic schizophrenic patients, we have elaborated this finding in subsequent reports [1–3, 5, 11]. Frequency of the septal spiking invariably related to the intensity of the psychotic symptoms. When psychotic

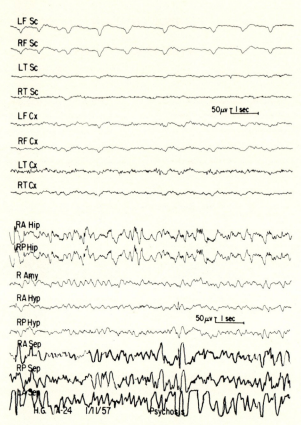

Fig. 5. Recordings from an epileptic patient during a psychotic episode. Note the high-amplitude spike and slow-wave activity has spread to involve the septal region, as well as the hippocampus.

signs and symptoms subsided, whether spontaneously or as the result of treatment, the spiking subsided (fig. 6). Concomitant slow-wave activity was often present. In schizophrenic patients, the abnormal pattern occasionally appeared in the hippocampus and amygdala as well as in the septal region, but less consistently. Recording changes were focal, not appearing in other deep leads and rarely appearing in recordings from the cerebral cortex or from the scalp.

In our series of 63 psychotic patients, 12 showed abnormalities in scalp recordings before electrode implantation, but had no history of seizures of any type. These patients were considered for operation because of their

Psychosis and Epilepsy

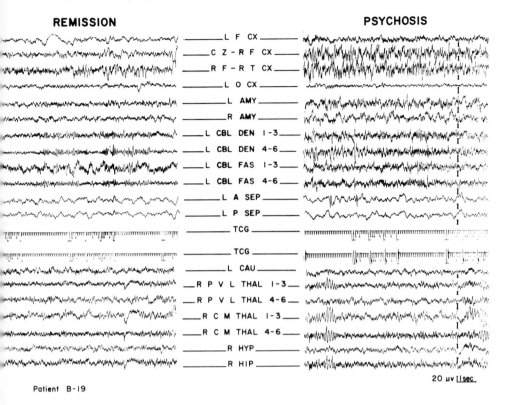

Fig. 6. Deep and surface recordings obtained from a schizophrenic patient. Note spiking in the septal leads during a psychotic episode.

intractable behavioral disorders. Many of them showed symptoms of extreme violence and aggression along with their psychotic behavior. The deep recordings of these patients when they were psychotic were in many ways similar to those obtained for schizophrenics and epileptics during their psychotic episodes. The abnormality was localized in the septal region, hippocampus, and amygdala. In association with extreme violence, recordings from the hippocampus and amygdala showed bursts of 12–18/s frequency spindling similar to patterns seen in *non*psychotic patients. But in addition there was increased spiking that roughly correlated with the intensity of the clinical symptoms (fig. 7, 8).

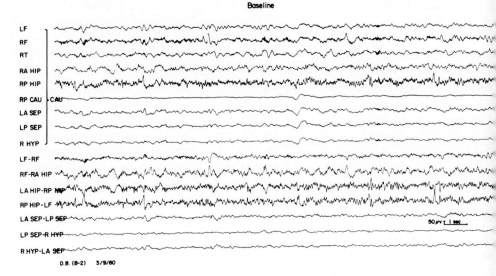

Fig. 7. Deep and surface recordings from a psychotic patient during a period of remission. Note occasional hippocampal sharp waves.

Comments

In our series of patients with deep electrodes, the correlation was consistent between septal spiking and psychotic behavior, regardless of the responsible pathologic process, that is, whether the psychosis was consequent to schizophrenia, epilepsy, or structural brain abnormality. Recordings in the deep temporal lobe nuclei also showed consistent changes during the patients' violent aggressive behavior, regardless of the initiating pathologic process. We have previously reported the appearance of recording changes with rageful or violent behavior induced by psychological factors [7, 11, 12]. For example, recall of anger-provoking incidents during a psychiatric interview prompted fast spindling in recordings from the septal region and amygdala.

Recording changes during epileptic seizures and in association with disturbed behavior occurred at the same focal anatomic sites. Thus, although the clinical symptoms are profoundly different, the anatomic substrate was the same. Since the nature of the recording changes reflecting activity of the cells is different in schizophrenia and epilepsy, it is logical

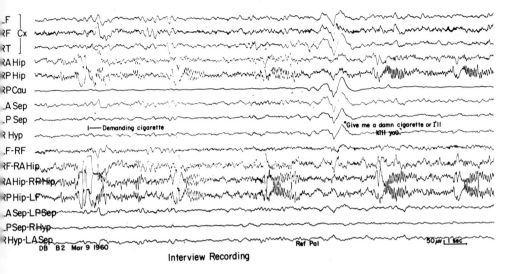

Fig. 8. Deep and surface recordings from the same psychotic patient as in figure 7 obtained during an interview when he was expressing intense rage. Note increased spiking associated with fast spindling in hippocampal leads and an increase in cortical slow-wave activity.

to hypothesize differences in the biochemical substrate of these cell populations. Moreover, since no cellular abnormality is consistently demonstrable in either schizophrenia or epilepsy, it seems reasonable to postulate that both disorders are metabolic but involve different biochemical lesions. Supportive of this hypothesis are recordings we have obtained during intense pleasure. During the explosive pleasure of orgasm, recordings from the septal region resemble those occurring during epileptic seizures, despite the sharply different subjective and objective clinical pictures [4]. Our recording equipment is obviously crude, merely showing that something is happening at a particular site while offering few clues as to the precise nature of the process. Light microscopy offers still less evidence of the nature of the ongoing process.

Limited experiments involving introduction of conventional cholinergic and aminergic transmitters, as well as peptide molecules, into these precise deep brain sites have provided little additional information concerning the nature of the metabolic process [3, 6]. The positron emission scanner, while promising, has thus far shown only what has been demonstrated by deep electrode techniques over the past 30 years, namely, that there is a metabolic abnormality in the basal frontal region (septal region) in association with psychosis and, further, that it spreads to become more generalized in epilepsy.

References

1. Heath, R.G.: Physiological and biochemical studies in schizophrenia with emphasis on mind-brain relationships. Int. Rev. Neurobiol. *1:* 299–331 (1959).
2. Heath, R.G.: Developments toward new physiologic treatments in psychiatry. J. Neuropsychiat. *5:* 318–331 (1964).
3. Heath, R.G.: Pleasure response of human subjects to direct stimulation of the brain: physiologic and psychodynamic considerations; in Heath, The role of pleasure in behavior, pp. 219–243 (Hoeber Medical Division, New York 1964).
4. Heath, R.G.: Pleasure and brain activity in man: deep and surface electroencephalograms during orgasm. J. nerv. ment. Dis. *154:* 3–18 (1972)
5. Heath, R.G.: Brain function and behavior: emotion and sensory phenomena in psychotic patients and in experimental animals. J. nerv. ment. Dis. *160:* 159–175 (1975).
6. Heath, R.G.; DeBalbian Verster, F.: Effects of chemical stimulation to discrete brain areas. Am. J. Psychiat. *117:* 980–990 (1961).
7. Heath, R.G.; Gallant, D.M.: Activity of the human brain during emotional thought; in Heath, The role of pleasure in behavior, pp. 83–106 (Hoeber Medical Division, New York, 1964).
8. Heath, R.G.; John, S.B.; Fontana, C.J.: The pleasure response: studies by stereotaxic technics in patients; in Kline, Laska, Computers and electronic devices in psychiatry, pp. 178–180. (Grune & Stratton, New York, 1968).
9. Heath, R.G.; John, S.B.; Fontana, C.J.: Stereotaxic implantation of electrodes in the human brain: a method for long-term study and treatment. IEEE Trans. biomed. Engng. *23:* 296–304 (1976).
10. Heath, R.G.; Mickle, W.A.: Evaluation of seven years experience with depth electrode studies in human patients; in Ramey, O'Doherty, Electrical studies on the unanesthetized brain, pp. 214–247 (Hoeber, New York, 1960).
11. Heath, R.G. and Tulane University Department of Psychiatry and Neurology: Studies in schizophrenia (Harvard University Press, Cambridge, 1954).
12. Lesse, H.; Heath, R.G.; Mickle, W.A.; Monroe, R.R.; Miller, W.H.: Rhinencephalic activity during thought. J. nerv. ment. Dis. *122:* 433 (1955).

R.G. Heath, MD, Tulane University School of Medicine, 1430 Tulane Avenue, New Orleans, LA 70112 (USA)

Carbamazepine, Temporal Lobe Epilepsy, and Manic-Depressive Illness

Robert M. Post, Thomas W. Uhde, James C. Ballenger, William E. Bunney Jr.

National Institute of Health, Bethesda, Md., USA

Introduction

Carbamazepine, increasingly the drug of choice for the treatment of temporal lobe epilepsy, may provide a unique focus for the study of the relationship of disturbances in temporal lobe function to behavioral pathology in the mood disorders. Carbamazepine may also serve as an important bridging drug between neurology and psychiatry. Preliminary data suggest that the behavioral and mood disturbances associated with both temporal lobe epilepsy [27] and manic-depressive psychosis are responsive to treatment with carbamazepine [7, 9, 85, 99, 101, 104, 113].

In this chapter we discuss the theoretical and clinical observations that led to our initiating the first double-blind, placebo-controlled clinical trials of carbamazepine's in depression and mania and the first clinical studies of the drug outside of Japan. We update the current status of our clinical trials of carbamazepine and discuss the implications of carbamazepine's efficacy in both epilepsy and affective illness.

The perspective that temporal lobe and limbic dysfunction may occur in affective illness, possibly through a kindling-like sensitization process, was central to the study of carbamazepine. Following our initial evidence of carbamazepine's effectiveness in some patients with affective illness, we were interested in pursuing whether the relatively greater effects of carbamazepine on the temporal lobe, in comparison with diphenylhydantoin, might be related to its psychotropic efficacy, or whether anticonvulsants in general might be of benefit in some patients. A double-blind crossover trial of diphenylhydantoin and valproic acid was therefore instituted.

We discuss the theoretical implications that valproic acid [36], in addition to carbamazepine, may also have positive psychotropic effects in affective illness, even though both agents have been reported to exacerbate symptoms in schizophrenics (see below). The possible implications of this differential effect on manic-depressive illness and schizophrenia will be discussed.

Attempts to assess the effects of electroconvulsive shock (ECS) on limbic (amygdala) kindling have also led us to the conceptualization of a different potential mechanism of action of electroconvulsive therapy – it too may be acting as a limbic anticonvulsant [114]. The discussion will thus focus on the clinical and theoretical implications of the relationship between convulsive and nonconvulsive disorders of the limbic system and disorders of affect and their treatment. Carbamazepine may prove to be a useful adjunct or alternative treatment approach for affectively ill patients who do not respond well to lithium carbonate.

Clinical Trials with Carbamazepine in Affective Illness

Background and Rationale

A number of converging observations led us to initiate clinical trials of carbamazepine in affective illness.

(1) As reviewed by *Dalby* [27], a substantial number of patients treated with carbamazepine for temporal lobe seizures were noted to have improvement in mood and behavior concomitant with improvement of their seizure disorder. In some instances investigators felt that the positive psychotropic effects of carbamazepine were demonstrated independent of either its anticonvulsant efficacy or the fact that it was substituted for other anticonvulsant agents which may have been causing side effects. Notable in these reports in epileptics was the improvement in motor and affective symptomatology, i.e., those areas of disturbance most commonly disordered in affective illness.

(2) Carbamazepine has relatively greater temporal lobe and limbic anticonvulsant effects than other agents (such as diphenylhydantoin) based upon both clinical and animal model studies [3, 5, 147]. The temporal lobe-limbic substrate has long been hypothesized to be involved in the regulation of affect [49, 59, 60, 73, 74, 76, 89, 130]. *Stevens* [this volume] discussed the risk factors predisposing individuals to the development of psychopathology, and *Sherwin* and *Trimble and Perez* discussed the relationship between epilepsy and schizophreniform psychosis. However,

relatively less theoretical attention has been directed to the finding of prominent affective disturbance in patients with temporal lobe epilepsy [47, 48, 75]. In many of the series of patients with temporal lobe epilepsy, affective disturbances have been reported in addition to those more typically presenting schizophreniform psychoses. In 536 interictal psychotic episodes, *Dongier* [33] found that 30% were of an affective nature with changes ranging from rapid mood swings to classic depressions and manias. *Flor-Henry* [41, 42] reported that affective disorders resembling manic-depressive psychosis occurred more often in patients with right compared to left temporal foci. *Taylor* (137) observed an increased incidence of depressive and manic-depressive-like symptoms with right-sided lesions. *Dalby* [26, 27] reported patients experiencing an increased incidence of depressive neurosis with right-sided lesions.

Conversely, *Flor-Henry* [41, 42] reported that patients with foci in the left temporal lobes presented with more schizophreniform psychosis resembling schizophrenia. These data are in part compatible with those of *Sherwin* [126 and this volume] and *Trimble and Perez* [this volume] who reported an increased evidence of psychosis with left temporal lobe foci. The study of *Bear and Fedio* [11] is also interesting in relation to affective symptomatology. Patients with right foci showed significantly more elation and increased ratings of affective disturbance compared to those with left foci. Comparing self-ratings and observer ratings the patients with right-sided foci also tended to use more denial and tended to enhance their own image relative to the observer, a 'polishing' tendency. In contrast, those with the left foci tended to exaggerate their negative traits, manifesting a 'tarnishing' self-perspective.

We repeatedly have observed a close parallelism among many of the symptoms reported to be associated with the interictal mood and behavioral disturbances of temporal lobe epilepsy and those observed in manic-depressive illness. There is a marked overlap in areas of symptomatology with one notable area of exception – prominent alterations in psychomotor activity. The increases in mania and the motor retardation and/or agitation with depression which occur in manic-depressive illness are not systematically reported in the epileptic patients. As discussed below, this overlap of clinical phenomenology might suggest a partial but incomplete overlap in underlying neural substrates. In addition, the lack of parallelism in relation to the activation-retardation spectrum may provide important clues regarding possible differences in underlying neurotransmitter alterations, particularly those involving catecholamine systems which have

Table I. Electrical kindling: major characteristics

1 Repeated stimulations
  a Initially subseizure threshold
  b Intermittent
2 Local after-discharges and seizure activity
  a Increases in amplitude, frequency
  b Increase in duration
  c Increase in complexity of wave form
  d Increase in anatomical spread
3 Replicable sequence of seizure stages
  Behavioral arrest, blinking and masticatory movements, head nodding, opisthotonus, contralateral then bilateral forelimb clonus, rearing and falling
4 Discharges kindle in quantum jumps
5 Limbic system kindles more readily than cortex
6 In kindled animals the history of convulsion development is recapitulated as seizure builds
7 Transfer effects to secondary sites; kindling facilitated in other sites even after primary site destroyed
8 Interference: a secondary kindled site interferes with primary site rekindling
9 No toxic or neuropathological changes evident; kindling is a transsynaptic process
10 Relatively permanent change in connectivity; a kindled animal will still seize after a 1-year seizure-free interval
11 Seizure may develop spontaneously in chronically kindled animals
12 After-discharges and seizures may oscillate or cycle in severity and duration

See *Goddard* et al. [50] regarding 1, 4–10; *Wada and Sato* [145] regarding 2–4, 6; *Wada* et al. [146] regarding 10–12; *Racine* [123]; *Pinel and Rovner* [92, 93]; *Post* et al. [116] regarding 12.

been closely linked to the regulation of psychomotor function. Thus, these initial observations helped suggest that carbamazepine, which can suppress abnormal excitability of temporal lobe and limbic structures, might be useful in the primary disorders of affective regulation, in addition to its utility in the treatment of temporal lobe epilepsy.

(3) Another line of evidence that suggested the possible utility of carbamazepine in affective illness derived from our formulation of a kindling-like sensitization phenomenon which may occur in the course of temporal lobe epilepsy and affective illness [97, 98, 110]. Carbamazepine has been demonstrated to inhibit the development of amygdala kindling in severel animal species [3–6, 144, 147]. In kindling, intermittent stimu-

lation of the amygdala or other brain areas is associated with progressive increases in after-discharge duration, complexity, and spread, and leads to the development of major motor seizures to a previously subconvulsant stimulation (table I) [50, 51, 123]. Kindling not only leads to the evolution of seizures but also produces a variety of long-lasting behavioral changes in animals (table II). As such, we have suggested that kindling may be a useful model for conceptualizing the evolution of behavioral pathology in patients with temporal lobe epilepsy [97, 110]. Repeated endogenous stimulation of the temporal lobe-limbic system might, as in exogenous electrophysiological kindling, lead to progressive augmentation of electrical responses in related pathways that either directly or indirectly modulate emotional and cognitive behavior. Dysregulated emotional behavior could result from the pathways' being either hypofunctional or activated and thus acting as a lesion directly [110], or by functionally disconnecting critical areas of emotional integration. If affective illness were associated with dysfunction or sensitization of temporal lobe limbic substrates involved in the regulation of affect, then carbamazepine might be postulated to be useful in the treatment of this process [102].

A variety of behavioral alterations do appear to show a behavioral sensitization or kindling-like progressive increase with the development of affective illness. In particular, across most studies since *Kraepelin* [65], recurrent affective illness, both unipolar and bipolar, tends to run a course of increasing frequency of cycles with a decreasing well interval between affective episodes [2, 5, 56, 57, 65]. In addition, patients with increased numbers of prior affective episodes appear to have more rapid or fulminant onsets of individual affective episodes [102]. While a variety of mechanisms may underlie the progressive development of psychopathology as a function of increased numbers of prior affective episodes, the behavioral sensitization and kindling models [98, 110] provide new areas for potential exploration of the role of anticonvulsant and other treatments.

(4) The early clinical trials of *Okuma* et al. [87] in the Japanese literature – the work of *Takezaki and Hanaoka* [135] was not known to us when we initiated our studies – gave added impetus to our clinical trials. These workers observed both acute and prophylactic response to carbamazepine during open clinical trials on the drug when it was used alone or added to previously ineffective regimens which included tricyclic antidepressants, lithium carbonate, or neuroleptic treatment. Their subsequent blind studies have also supported an antimanic and prophylactic effect of carbamazepine [85, 86].

Table II. Behavioral effects of electrical kindling: long-lasting spontaneous and drug-induced alterations

Effect	Investigator(s)
Amygdala kindling	
Decreased rat killing (cat)	*Adamec* [1]
Decreased spontaneous rearing and decreased cocaine-induced rearing (rat)	*Post* et al. [117]
Decreased amphetamine-induced hyperactivity (rat)	*Ehlers et al. [34]*
Decreased exploration[1] (rat)	*Weingartner and White* [150]
Weight gain (rat)	*Innes* et al. [62]
Increased aggression (rat) (not replicated)	*Pinel* et al. [94] *Bawden and Racine* [10]
Increased predatory response (rat)	*McIntyre* [78]
Increased rage reaction[1] (cat)	*Yoshii and Yamaguchi* [152] *Gunne and Reis* [58]
Increased play, decreases and increases in aggression[1] (monkey)	*Deflorida and Delgado* [28]
Inhibitory avoidance (rat)	*Boast and McIntyre* [18]
Decreased conditioned emotional response (reversed by adrenalectomy in rat)	*McIntyre* [78] *McIntyre and Molino* [79]
Intense fear[1] (man)	*Heath* et al. [60]
Mood change and pressure of speech[1] (man)	*Ervin* et al. [37] *Stevens* et al. [133] *Goddard and Morrell* [51]
Nucleus accumbens kindling	
Decreased amphetamine-induced hyperactivity (rat)	*Ehlers* et al. [34]
Hippocampal kindling	
Increased aggression (rat)	*Pinel* et al. [94]
Lateralized state-dependent learning (rat)	*Stokes and McIntyre* [134]
Ventral tegmentum$^{(A_{10})}$ kindling	
Increased bizarre behavior and decreased social activity (cat)	*Stevens and Livermore* [132]
Hypothalamus kindling	
Increase in food intake[1] (cat)	*Delgado and Anand* [29]
Increase in food or water intake [1] (rat)	*Valenstein* et al. [143]
Lower eating and drinking thresholds[1] (rat)	*Wise* [151]

[1] Not typical kindling paradigm

Thus, based on these diverse clinical, theoretical, and empirical observations focused around the issues of kindling and limbic system dysfunction in affective illness [96, 110], we initiated our studies of carbamazepine in 1975.

Methods

Patients were admitted to the 3-West Clinical Research Unit of the Section on Psychobiology of the Biological Psychiatry Branch of the National Institute of Mental Health for treatment of affective illness of an incapacitating nature severe enough to require hospitalization. In most instances patients met Research Diagnostic Criteria and DSM III criteria for major affective illness. Two patients with depressive symptomatology associated with borderline personality disturbance and one case with multiple personality disorder were included in the clinical trial. In each instance patients received a thorough history, and neurological and laboratory examinations which were negative for evidence of seizure disorder. Routine EEGs, lumbar punctures, skull series and, in most instances, CAT scans were within normal limits. In several patients where an underlying seizure disorder was suspected, more detailed EEG studies with sleep deprivation and/or use of nasopharyngeal leads were instituted and were also negative. Thus, it would appear from routine neurological and electrophysiological evaluation that the affectively ill patients included in this series were not suffering from clinically diagnosable seizure disorders. Moreover, their presentations of affective symptomatology were highly similar to other patients on our psychiatric unit who responded to the classical antidepressant and antimanic agents, including tricyclic antidepressants and lithium carbonate.

Patients were evaluated longitudinally using double-blind, placebo-controlled methodology [7, 9]. Patients received pink opaque capsules with placebo/active compounds throughout the majority of their hospital course. The nurses who performed clinical ratings were unaware of whether patients were on placebo or active compounds, or which active compounds the patients might be receiving at any given time. Patients were rated twice daily by consensus of trained nursing observers in global measures of mania, depression, psychosis, anger, and anxiety [22]. Three times weekly brief psychiatric ratings (BPRS) were performed by the same nursing staff blind to the nature of the clinical intervention [88]. In addition, patients, when clinically able, rated themselves once or twice daily on a number of rating scales and a side effect checklist to further ascertain a subjective sense of clinical response. These instruments have previously been utilized in the evaluation of lithium carbonate [54], piribedil [107], pimozide [108], and tricyclic and monoamine oxidase inhibitor antidepressant drugs [81, 82]. Patients entering into the carbamazepine study were not systematically crossed over to treatment with these other psychotropic agents. However, for comparison with a similarly screened and diagnosed patient population, the degree of clinical improvement on carbamazepine will be compared to that achieved with these other psychotropic agents.

Treatment with carbamazepine was instituted at initial doses of usually 200–400 mg/day and increased gradually to an end-point of clinical efficacy, dose-limiting side effects, or initially, blood levels in the range of 8–12 µg/ml. After 1980, following lack of evidence of a relationship between carbamazepine blood levels and clinical response (see below), the initial blood level range was no longer utilized as the key indicator of dose

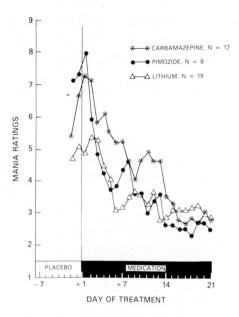

Fig. 1. The onset of antimanic effects of carbamazepine following double-blind substitution for placebo is roughly comparable to that achieved by pimozide or lithium carbonate in similarly diagnosed and rated patients.

increases. The dose was gradually increased and titrated against side effects or clinical response to a limit of 1,600 mg/day (increased to 2,200 mg in 1981). Several patients have been treated at doses of 1,600–2,000 mg/day without problematic side effects [91].

Blood and CSF levels of carbamazepine and carbamazepine-10, 11-epoxide were also assessed using gas chromatographic or high pressure liquid chromatographic (HPLC) techniques [15]. Routine clinical chemistries were obtained and white counts and platelets were determined at weekly intervals during the initial phases of the clinical trial.

Lumbar punctures were performed using standardized techniques [55] during a medication-free interval of at least 2 weeks and during treatment with the maximum effective dose of carbamazepine approximately 4–6 weeks into the clinical trial [99, 113]. A variety of other clinical and biological measures were obtained before and after carbamazepine administration and will only be briefly presented here. A more detailed discussion of the physiology, biochemistry, and pharmacology of carbamazepine is presented elsewhere [103].

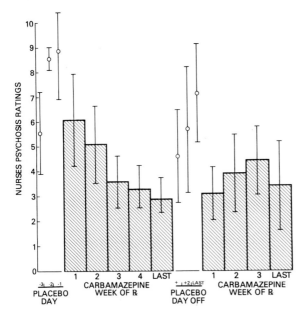

Fig. 2. Improvement in global psychosis ratings occurs with the blind initiation of carbamazepine treatment in 4 manic patients who completed this 'off-on-off-on' design. Following placebo substitution, psychosis reemerges even by the first and second day off carbamazepine; reinstitution of active treatment is again associated with an antimanic and antipsychotic response.

Results

Acute and Prophylactic Antimanic Efficacy of Carbamazepine

7 of 12 manic patients showed a partial to good response to carbamazepine; 5 demonstrated a good response and 4 showed marked relapses following placebo substitution. The overall comparative time course of the antimanic effects of carbamazepine compared to lithium carbonate or the neuroleptic pimozide are illustrated in figure 1. The time course of antimanic response in these 12 patients appears roughly similar to that achieved with the relatively specific dopamine receptor blocking agent pimozide or to that achieved by the classical treatment of affective illness with lithium carbonate. In 4 of the manic patients, observations of carbamazepine's efficacy and relapse following placebo substitution was repeat-

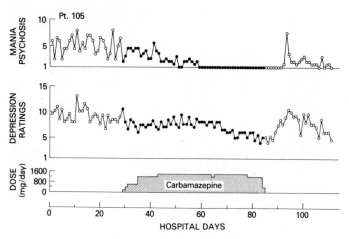

Fig. 3. Active treatment with carbamazepine is associated first with an antimanic, then with an antidepressant response in this bipolar patient. Following blind discontinuation of carbamazepine, a rapid increase in depression is evident.

edly documented in an 'off-on-off-on'design (fig. 2). As illustrated in figure 3, improvement may also be noted in depressive and dysphoric symptoms accompanying or following manic episodes. This 53-year-old patient was also hospitalized for prolonged periods of time prior to the NIMH in a state institution and was nonresponsive to lithium, neuroleptic, and tricyclic antidepressant treatment. Following the onset of a prolonged manic episode, carbamazepine treatment was instituted on a double-blind basis. Manic ratings improved rapidly followed by a decrease in depression ratings as well (fig. 3). Following double-blind placebo substitution, a rapid increase in depressive symptoms was noted, accompanied by a transient reappearance of manic symptomatology. This patient then continued in a prolonged inpatient treatment phase with carbamazepine and responded sufficiently well such that she was able to be discharged to an outpatient treatment setting for the first time in 22 years. She was maintained on carbamazepine alone and did well except for two manic episodes at times of reported dose reduction and/or discontinuation. During these episodes the patient was treated with neuroleptics in another institution without substantial clinical effect and did not respond until carbamazepine treatment was reinstituted. The overall prophylactic response to carbamazepine is illustrated in figure 4.

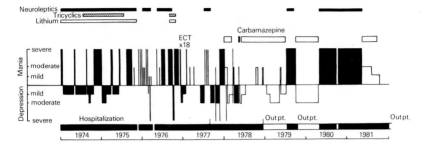

Fig. 4. A partial prophylactic response to carbamazepine is illustrated in a patient who had been hospitalized in another institution since 1955 with rapidly cycling manic-depressive illness. She was unresponsive to a variety of psychotropic agents, including lithium and neuroleptics, and had only a transient response to ECT. The patient, whose acute response is illustrated in figure 3, became well enough on maintenance carbamazepine to be discharged from NIMH. She lived independently as an outpatient for approximately 9 months on two occasions until she was rehospitalized for manic episodes which did not remit until carbamazepine treatment was substituted for neuroleptics.

As illustrated in figure 5, a 55-year-old patient (No. 167) with a history of continuous hospitalization in a state institution since 1954, non-responsive to a variety of treatment modalities including neuroleptics, lithium, and electroconvulsive therapy was admitted to NIMH and, following several mood cycles observed on placebo, was successfully treated with carbamazepine.

Comparison of Carbamazepine and Diphenylhydantoin

This patient is of particular interest not only because of her well-documented dose-related responsiveness to carbamazepine (fig. 6), but also because of our initial observations when the patient was switched on a double-blind basis to treatment with another anticonvulsant – diphenylhydantoin. Although the patient showed rapid onsets of improvement during two separate clinical trials on carbamazepine, following treatment with diphenylhydantoin at doses of 600 mg/day, no evidence of antimanic or antipsychotic efficacy was observed (fig. 7). These preliminary observations in a well-documented carbamazepine responder suggest that affectively ill patients may be selectively responsive to anticonvulsant drugs and may not respond to anticonvulsants as a class, as typified by the lack of response to diphenylhydantoin in contrast to carbamazepine in our first

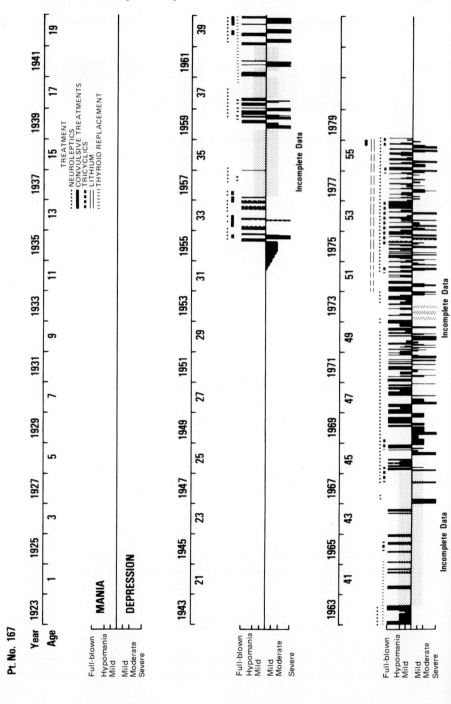

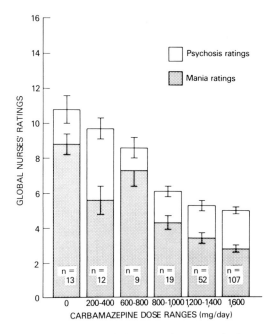

Fig. 6. Improvement in both manic and psychotic components of the illness occurred in a dose-dependent fashion during treatment of this manic patient with carbamazepine.

patient in this double-blind crossover study. Based on the initial positive observations of *Emrich* et al. [36], we plan to further assess the efficacy of the anticonvulsant valproic acid in this patient and in other treatment-resistant, rapidly cycling manic-depressive patients.

Prophylactic response in this patient, as illustrated in figures 6 and 7, is further documented in table III. During treatment with carbamazepine compared to placebo, there was a 85 and 25% reduction in the number of days of severe depression and mania, respectively, calculated on a days ill per year basis, as well as a 129% increase in the number of improved days during carbamazepine treatment compared to that observed during placebo.

Fig. 5. The pattern of chronic and rapidly cycling illness is illustrated in a life chart of a patient whose acute and dose-dependent response to carbamazepine is illustrated in figures 6 and 7. This patient had also been unresponsive to lithium and neuroleptic treatment. The patient's double-blind clinical trial of carbamazepine prophylaxis is summarized in table III.

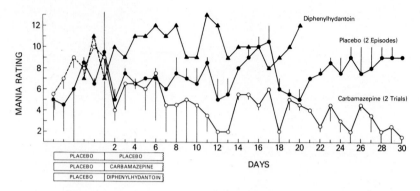

Fig. 7. Carbamazepine treatment of two manic episodes resulted in improvement in mania ratings compared to two untreated episodes. Treatment with phenytoin (600 mg/day peak dose) resulted in no evident improvement. This patient also failed to respond to crossover to valproic acid (doses up to 2.5 g/day). All ratings were performed double-blind.

Table III. Evidence of effective carbamazepine prophylaxis compared to placebo in an extended clinical trial in a manic-depressive patient

Clinical state	Treatment				
	placebo (263 days)		carbamazepine (507 days)[1]		
	days	days/year	days	days/year	% reduction in days of severe illness
Severe depression[2]	16	105.5	23	16.3	84.5
Severe Mania[3]	116	161.0	174	123.1	24.5
					% increase in improved days
Improved	71	98.5	319	225.6	129

[1] Days when patient was treated with carbamazepine ≥ 800 mg/day.
[2] Depression = ratings ≥ 7, no mania.
[3] Mania = ratings ≥ 5.

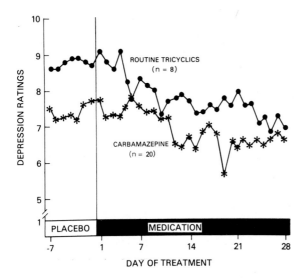

Fig. 8. A rough parallelism in the time course of improvement in depression is noted during treatment with carbamazepine or the routine tricyclics, imipramine or amitriptyline. The carbamazepine-treated patients with the highest initial severity of depression responded better than those who were initially less depressed.

Antidepressant Response

The time course of antidepressant response to carbamazepine is illustrated in comparison to that achieved with the standard tricyclic antidepressants, imipramine and amitriptyline in figure 8. Again, the rough parallelism of time course and magnitude of antidepressant effect is noted, although a randomized or crossover design was not possible in this phase of the study. Of the first 25 depressed patients to enter this study to date, 48% showed a good or moderate response, while 24% showed a poor or partial response, and 28% showed mild worsening in depressive symptomatology on carbamazepine. It is of interest that carbamazepine responders had significantly increased levels of severity of depression before beginning active treatment compared to the nonresponders.

It is also noteworthy that of the 8 best responders, 6 showed evidence of an increase in depressive symptoms in either of the 2 weeks following placebo substitution. In contrast, none of the 7 patients who became slightly more symptomatic during the carbamazepine trial showed an exacerbation of symptomatology in the same period following carbamazepine withdrawal (chi square=8.75, p<0.01). These data support our clinical impression that discontinuation of carbamazepine is not associated with a withdrawal syndrome per se, but rather that, in some instances, patients who improve can show a rapid relapse toward their initial degree of affective symptomatology when the drug is discontinued (fig. 2, 3). However, in the nonresponders there is little evidence of a nonspecific withdrawal syndrome. Moreover, when discontinuation of active compound is associated with relapse, the original symptomatology gradually emerges and there is little evidence of the development of an organic brain syndrome or other toxic manifestations that are usually indicative of a withdrawal reaction. These data are also consistent with those derived from clinical and experimental studies of carbamazepine withdrawal in epilepsy. These studies suggest that the usual frequency of seizures, but not an increased frequency, may appear following discontinuation of treatment with carbamazepine.

Thus, the appearance of both improvement of affective symptomatology associated with active treatment and exacerbation following its discontinuation provide evidence of carbamazepine efficacy in at least a subgroup of affectively ill patients. The majority of manic patients in our study were previously noted to be nonresponsive to treatment with lithium carbonate. The apparently acute and prophylactic effects of carbamazepine in both the manic and depressive phases of the illness thus suggest the possibility of its utility as an alternative treatment approach for some patients with affective illness. Carbamazepine may also be important because, in contrast to lithium carbonate, it does not cause (but actually reverses) the diabetes insipidus syndrome (see below).

Improvement in Sleep and the Occurrence of Side Effects

Treatment with carbamazepine was associated with rapid improvement in sleep, noted both by nurses performing half-hourly sleep checks and subjectively by the patients. In the first 22 patients studied, improvement in sleep was documented during the first week of carbamazepine treatment of either the manic or the depressive phase of the illness by this half-hourly sleep check criteria (fig. 9). In addition, preliminary studies, conducted in collaboration with *Gillin* using EEG monitored sleep techniques, suggest the increase in sleep occurred largely in slow wave sleep components (stages 3 and 4). It is also of interest that this improvement in sleep was occurring without notable increases in daytime sedation in the

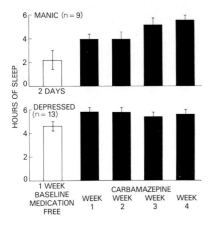

Fig. 9. In patients studied utilizing half-hourly sleep checks by nurses who were unaware of the medication status. Rapid improvement in sleep was noted in the first week of treatment with carbamazepine in both the manic and depressive phases of the illness.

majority of patients. While some patients experience some sedation and drowsiness, this is usually dose-related, occurs early in the clinical trial, and tends to disappear with maintenance of the same dose or after transient dose reductions. We have not observed extrapyramidal side effects or Parkinsonian motor inhibition as is often seen following neuroleptics. *Stevens* [personal communication, June 1981] has observed two severe dystonic reactions in schizophrenic patients when carbamazepine was added to high-dose neuroleptic treatment. *Martinon* et al. [77] reported dystonic reactions in 4 children treated with carbamazepine. Our clinical observations of the lack of Parkinsonian side effects are consistent with the lack of increase in the dopamine metabolite homovanillic acid (HVA) in cerebrospinal fluid during carbamazepine treatment [101, 103], and suggests that carbamazepine is not acting like the classical neuroleptics by blocking dopamine receptors.

Other side effects tended to be those usually associated with too rapid increases in anticonvulsant dose and included dizziness, dysarthria, and a subjective sense of clumsiness [9]. These also appeared to be dose-related and most individuals did not find the side effects to be problematic. 3 of the first 37 patients treated with carbamazepine required discontinuation of the drug because of the development of a macular, sometimes pruritic

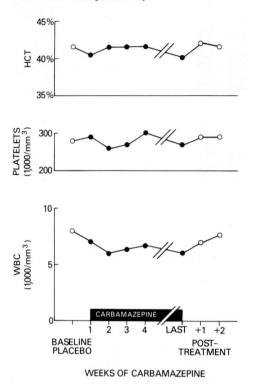

Fig. 10. Hematologic indices during carbamazepine treatment. In the first 22 patients studied, mild decreases in white blood cell count were observed which returned to baseline following discontinuation of carbamazepine. Platelet count was not significantly affected.

rash. In only 1 patient was carbamazepine discontinued because of a white count decrease below 3,700 white cells per cubic millimeter. However, a statistically significant but clinically insignificant reduction in the blood count was observed across the patient population during carbamazepine treatment with an increase to pretreatment levels following carbamazepine withdrawal (fig. 10). These data in a relatively small number of patients are consistent with those in the much larger recent neurological literature that the increasing use of carbamazepine in the treatment of epilepsy has not been associated with increased reports of the development of aplastic anemia [*Scoville,* personal commun.], as had been reported as a rare side effect in the literature of a decade ago [95].

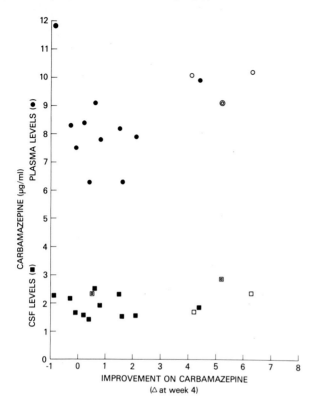

Fig. 11. Degree of clinical antidepressant (shaded symbols) or antimanic (open symbols) response assessed at week four or at the time of the lumbar puncture was not significantly related to plasma or CSF levels of carbamazepine.

Table IV. Plasma and CSF levels of carbamazepine and its 10,11-epoxide metabolite in patients with affective illness

	Carbamazepine, µg/ml	Carbamazepine-10,11-epoxide, µg/ml	Epoxide, %
CSF (n = 16)	2.01 ± 0.10	0.82 ± 0.08	41.0 ± 3.5
Plasma (n = 16)	8.53 ± 0.39		
Unbound carbamazepine (CSF/plasma × 100)	23.6%		

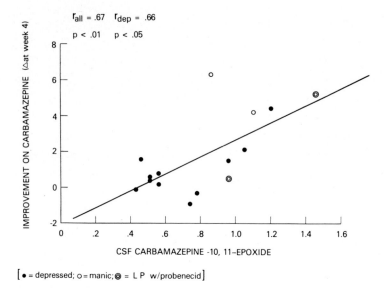

[● = depressed; ○ = manic; ◎ = L P w/probenecid]

Fig. 12. The degree of clinical improvement on carbamazepine was significantly correlated with CSF levels of carbamazepine-10, 11-epoxide in 11 depressed patients (r = 0.66) or in the combined group of 15 depressed and manic patients (r = 0.67).

Blood and CSF Levels of Carbamazepine and Carbamazepine-10, 11-Epoxide: Relationship to Clinical Response

Doses and blood levels of carbamazepine in our affectively ill patients were in a range similar to those used in epileptic patients. We also measured CSF levels of carbamazepine and its 10, 11-epoxide (table IV). In the initial group of patients studied, blood and CSF levels of carbamazepine were not correlated with degree of clinical antimanic or antidepressant response (fig. 11). However, CSF levels of carbamazepine-10, 11-epoxide (fig. 12) were significantly correlated (r = 0.67, n = 15, p<0.01) with clinical response measured at week 4 (or at time of LP). Carbamazepine-10, 11-epoxide has anticonvulsant effects in animals [38, 40, 45], but has not yet been given in clinical trials in epilepsy. Our preliminary data suggest that the 10, 11-epoxide may possess active psychotropic effects which could, either alone or in addition to the parent compound, account for carbamazepine's therapeutic efficacy in affective dysregulation.

Carbamazepine's Effects on Classical Neurotransmitter Systems

The anticonvulsant efficacy of carbamazepine may be related, at least in part, to its effects on norepinephrine (NE) since 6-hydroxydopamine inhibits its anticonvulsant action [121, 122]. Carbamazepine blocks reuptake of NE similar to that of imipramine-like tricyclics, but also blocks stimulation-induced release of NE in the rabbit ear artery preparation [119]. The NE metabolite MHPG is decreased in CSF but not in urine in our patients treated with carbamazepine [112; *Gordon, Jimerson, Smith,* unpubl. observations]. CSF-NE was decreased during carbamazepine treatment in manic but not depressed patients [112; *Lake,* unpubl. data].

Probenecid-induced accumulations of HVA were significantly decreased during carbamazepine treatment, but baseline CSF levels of HVA and 5HIAA were not significantly changed [101]. Carbamazepine as well as other anticonvulsants have been reported to alter GABA turnover [12], a finding of particular interest in relation to GABA hypotheses not only of schizophrenic disorders but also of affective illness [13, 36]. Carbamazepine did not significantly affect GABA measured in CSF of our patients, however [99].

Probenecid-induced accumulation of c-AMP and c-GMP was decreased on carbamazepine when compared with medication-free values, but baseline levels were unaffected. Moreover, the degree of antidepressant response was inversely correlated with basal medication-free levels of c-GMP in CSF [106]. These findings are of interest in relation to theories of the role of cyclic nucleotides in both epilepsy and affective illness.

Peptide Functions

The effects of carbamazepine on central nervous system opioid function is of interest because of: (a) carbamazepine's effects in pain syndromes; (b) the postulated relationship between opioids and limbic seizures [61], and (c) carbamazepine's potentiation of opiate-induced running in mice [64]. Although we find no significant effect of carbamazepine on total opioid binding in CSF measured by a radioreceptor assay [113], this clearly does not rule out effects of carbamazepine in discrete opiate peptide systems or brain regions. We did observe a positive correlation between pretreatment levels of CSF opioid activity and degree of antidepressant response to carbamazepine (fig. 13).

Somatostatin in CSF was significantly decreased during treatment with carbamazepine, but not during treatment with the standard tricyclics imipramine and amitriptyline, and was increased by zimelidine [125].

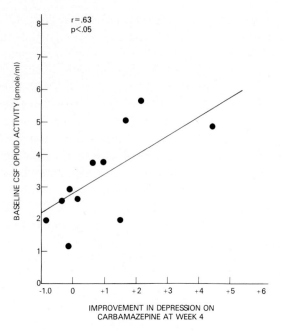

Fig. 13. Those patients with higher pretreatment levels of opioid binding activity in CSF showed a better antidepressant response to carbamazepine than those with lower initial values. Carbamazepine treatment itself did not significantly affect this measure of CSF opioid-binding activity.

These findings are of interest in relation to the rich innervation of the limbic system by somatostatin and the differential clinical spectrum of efficacy of carbamazepine compared to other antidepressants.

Carbamazepine also has an interesting clinical and biological profile of effects on vasopressin. It has been used to treat diabetes insipidus [129, 148], in contrast to lithium carbonate which induces a diabetes insipidus syndrome. Carbamazepine also alters the vasopressin response to a hypertonic saline load (decreases it because of its agonist properties) in a direction opposite to lithium [52]. It also produces a dose-related hyponatremia with the greatest effects observed on those with the lowest pretreatment serum sodium concentration [91, 142]. Not only are these effects of carbamazepine of clinical interest as it may be used as a substitute treatment for patients with lithium-related diabetes insipidus, but they raise the possibility that in addition to carbamazepine's positive effects on mood described here, its effects on cognition [139] and even seizures [120] could be partially related to vasopressinergic function.

The biochemical and physiological profile of carbamazepine in contrast to other anticonvulsants and other routine psychotropic agents such as lithium, tricyclics, and neuroleptics should provide important clues to its mechanisms and sites of action and possibly help to identify substrates altered in the disorders of affect regulation and in seizures.

Discussion

General Aspects
Our initial double-blind, placebo-controlled clinical trial suggests that some patients with primary affective disorders may respond both acutely and prophylactically to treatment with carbamazepine in either their depressed or manic phases of the illness. Although the number of patients studied to date is relatively small, the evidence in selected patients utilizing the 'off-on-off-on' design is convincing. That is, patients showed clinical improvement during treatment with active drug, showed relapse following placebo substitution, and then improved again following reintroduction of active treatment with carbamazepine. It is noteworthy that many of the best responders to carbamazepine had been prior failures or incomplete responders to lithium carbonate. However, 1 patient who showed unequivocal evidence of carbamazepine response [patient Nr. 1, fig, 1, in ref. 7] also showed a good prophylactic response to lithium. These observations are consistent with those of *Okuma et al.* [87] based on open clinical observations that carbamazepine might be useful in previous nonresponders to lithium carbonate, but that there was no specific predictive relationship between response or nonresponse to these two agents. More precise description of the relationship of carbamazepine and lithium responses within the same patients require further documentation in larger patient populations.

It remains to be determined whether carbamazepine responders and lithium carbonate responders might be differentiated on some clinical or biological measure. If there were subgroups based on differential response, it would be of obvious theoretical import in relation to possible differential pathophysiology of the illness; hypothetically, lithium carbonate might affect one underlying pathological process and carbamazepine another. Although lithium has effects on limbic excitability [30], it does not affect

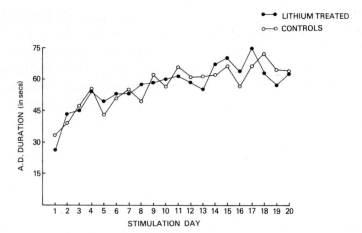

Fig. 14. Rate of growth of amygdala after-discharges in the rat following once-daily stimulation for 1 s with 60 Hz, 200 μA biphasic square waves was identical in lithium (n = 14) and plain diet controls (n = 18). Lithium carbonate (approximately 3 mEq/kg) was added to moistened powdered food in order to achieve chronic serum lithium levels in the rat between 0.2 and 1.2 mEq/l.

the rate of development of amygdala kindling, as measured by growth of after-discharges (fig. 14), as does carbamazepine. These and other differential physiological and biochemical effects of lithium and carbamazepine could be related to their different clinical spectra in temporal lobe epilepsy and affective illness. The relative efficacy of lithium and carbamazepine in mood and behavior disorders of temporal lobe epilepsy also remains fertile ground for further investigation.

We raise the possibility that the more potent limbic anticonvulsant properties of carbamazepine, compared to those of lithium [3], might be related to its utility in some treatment-resistant, rapidly cycling manic-depressive patients. We are not suggesting that these patients have seizures but that paroxysmal or dysregulated activity may be successfully modulated or dampened by carbamazepine, as appears to be the case in trigeminal neuralgia [43, 44]. It is also possible that responders may be distinguished by their longitudinal course of illness; milder forms of affective disorder appear responsive to both lithium and carbamazepine, but components of the illness associated with more progressive or rapid cycling, characteristic of many of our patients (fig. 5), may respond better to carbamazepine in some individuals.

From the clinical perspective, the successful treatment of a number of patients who were previously documented to be lithium nonresponders with the anticonvulsant carbamazepine raises the possibility that this agent may eventually have an important place in the therapeutic armamentarium for the treatment of some types of affective illness, particularly those nonresponsive to lithium carbonate. In this regard it is also noteworthy that many of the side effects of carbamazepine and lithium appear to be opposite in direction. For example, as noted above, lithium carbonate appears to impair the action of antidiuretic hormone (vasopressin), leading in a number of patients to the development of clinically substantial diabetes insipidus. In some patients this may be of problematic proportions, reaching a magnitude of 5–10 liters of urine output per day. In contrast, carbamazepine has been used in the treatment of diabetes insipidus of a variety of etiologies and appears to have direct or indirect vasopressin agonist effects [14, 52, 129, 138, 142, 148; *Berrettini and Post*, unpubl. data]. Thus, from a clinical perspective, those patients with severe lithium-induced diabetes insipidus may be appropriate for consideration of a trial on carbamazepine, particularly since preliminary reports indicate that carbamazepine may be useful in treating diabetes insipidus that persists following lithium discontinuation [*Brooks* et al., personal commun.]. Whether carbamazepine will reverse lithium-induced diabetes insipidus without the discontinuation of lithium carbonate remains for further study, although our unpublished observations and the data of *Ghose* [46] suggest that this may not be the case.

The apparently opposite effects of carbamazepine and lithium carbonate on vasopressin function may also have interesting implications for the mechanisms of action of these compounds, as well as associated central side effects. In particular, recent data suggest that vasopressin may improve learning and memory in a variety of animal and human learning paradigms [31, 32, 53, 149]. In addition, it has been increasingly recognized that cognitive impairment may be associated with chronic lithium treatment [124]. Whether or not this is related to its inhibitory effects on vasopressin function remains to be delineated but it raises the possibility that, because of carbamazepine's vasopressin agonist effects, it may not be associated with such cognitive impairment. The differential effects of lithium and carbamazepine on vasopressin remains an intriguing area for further exploration in relation to the mechanism of action of side effects and/or clinical efficacy. The apparently opposite effects of these two agents on vasopressin function do not necessarily rule out the possibility that they exert their therapeutic effects in affective illness in part through the vaso-ergic system.

It is also of clinical and theoretical interest that these two agents exert opposite effects on the white blood cell count. Lithium and carbamazepine, respectively, produce increases and decreases in the white blood cell count which are statistically significant, although usually clinically insubstantial.

Theoretical Implications for Substrates Underlying Disorders of Affective Regulation

At the present time only indirect evidence is available to suggest that carbamazepine may be exerting its positive psychotropic effects in affective illness through its temporal lobe-limbic antiexcitability effects. One line of evidence is the spectrum of clinical and experimental efficacy that carbamazepine possesses in the treatment of seizure disorders. Although carbamazepine is useful in the treatment of major motor seizures of a variety of etiologies, it is increasingly recognized as a treatment of choice for temporal lobe epilepsy [90]. In addition, the studies of *Albright and Burnham* [3] indicate that carbamazepine has the highest ratio of limbic to cortical anticonvulsive efficacy in the kindling seizure models. While these data suggest that carbamazepine has a better clinical and experimental spectrum of action in seizures related to the temporal lobe and limbic system compared to classical or cortical seizure disorders, it is possible that carbamazepine is exerting its positive psychotropic effects in affective illness through mechanisms unrelated to those of its anticonvulsant efficacy. Other strategies will be required in order to assess whether carbamazepine is acting at the level of the temporal lobe and limbic system.

One, albeit indirect, approach to the problem is the relative assessment of psychotropic effects of different anticonvulsant agents with different profiles of action. We are instituting clinical comparisons of carbamazepine with diphenylhydantoin (fig. 7) and valproic acid in order to pursue this strategy. This patient, who demonstrated an unequivocal positive response to carbamazepine, showed no evidence of a positive response to diphenylhydantoin or valproic acid. Further clinical trials are obviously in order, particularly in light of the early clinical reports of some positive effects of diphenylhydantoin [19, 20, 63, 141]. If carbamazepine and not diphenylhydantoin is documented to be effective, it would be consistent with the view that carbamazepine's greater efficacy in temporal lobe and limbic disorders, compared to diphenylhydantoin [3], could also be related to its efficacy in affective dysfunction. Study of the differential biochemical and physiological effects of carbamazepine and diphenylhydantoin would also prove useful in this regard, especially since these two anticonvulsants share many effects.

Initial studies of *Emrich* et al. [36] and *Lambert* et al. [66] suggest in both open and double-blind clinical trials that valproic acid or its derivative dipropylacetamide may show acute and prophylactic efficacy in the treatment of patients with manic and depressive illness, some of whom were also lithium nonresponders. These data raise provocative questions regarding the possible mechanisms of neurotransmitter effect of valproic acid, as well as the neuroanatomical locus of its efficacy. A variety of studies suggest that valproic acid may be increasing GABA function in the central nervous system by several different effects on GABA metabolism [13, 36]. Although it appears to have a different spectrum of clinical efficacy than some of the other anticonvulsant agents, including useful effects in the spike and wave disorders of petit mal, its usefulness in the treatment of temporal lobe and limbic disturbances remains to be more definitively assessed. It is noteworthy that valproic acid and a variety of other agents which potentiate GABA functon are useful anticonvulsants in the model of amygdala kindling [71, 72, 83, 84, 105, 123, 136]. A substantial body of indirect data also supports a role for GABA in the disorders of affect [13, 35, 36, 100].

Differential Effects of Carbamazepine and Valproic Acid in Affective Illness and Schizophrenia?

While *Emrich* et al. [36] have reported therapeutic effects of valproic acid in manic-depressive psychosis, *Lautin* et al. [70] have reported that valproic acid and other GABA agonists may exacerbate schizophrenic symptomatology. These data are parallel to those of carbamazepine. *Post* et al. [101, 103], *Ballenger and Post* [7, 9], *Silberman and Post* [127], and *Okuma* et al. [85, 87] have observed therapeutic effects of carbamazepine in patients with affective psychoses, including many severely affectively ill patients with sufficient psychotic symptoms to qualify for schizoaffective diagnoses by Research Diagnostic Criteria [128] in our series. *Stevens* et al. [131] have noted exacerbation of symptomatology in schizophrenics maintained on their neuroleptic compounds when carbamazepine therapy was added. It is of interest that carbamazepine may, in some instances, precipitate or exacerbate symptoms, while in others effectively treat schizophreniform psychosis associated with epilepsy [27] (see below). In our patients with affective illness, carbamazepine alone or in conjunction with neuroleptic treatment has been useful in the treatment of even psychotic symptoms associated with affective disorders (fig. 2). Should the initial observations that valproic acid and carbamazepine exacerbate schizophrenia, and our findings and those of *Emrich et al.* [36] of the therapeutic effects of these anticonvulsants in affectively ill patients be replicated, a clinical and theoretical tool might exist for further dissection of the possible substrates and mechanisms underlying the differential responsivities in these two major psychoses. Moreover, since the neuroleptic agents appear to be useful in the treatment of both manic and schizophrenic psychoses, a differential response in these two syndromes to the anticonvulsants carbamazepine and valproic acid would suggest actions through systems other than those related to dopamine metabolism. The common effects of carbamazepine and valproic acid might be explored in providing clues to the differential biology of manic and schizophrenic psychoses.

The Paradox of Electroconvulsive Seizures and an Anticonvulsant Both Displaying Therapeutic Efficacy in Primary Affective Illness

A provocative theoretical issue is raised by the initial studies suggesting the efficacy of the anticonvulsants carbamazepine and valproic acid in affective illness when compared with the well-known efficacy of electroconvulsive therapy [39]. How might one conceptualize that seizures themselves and agents which prevent seizures are both efficacious in the treatment of the same behavioral disturbance? Among many potential explanations, we might suggest two possibilities. First, electroconvulsive therapy may be acting itself as an anticonvulsant. Preliminary data in our laboratory in collaboration with *Putnam* and *Contel* indicate that electroconvulsive seizures (ECS) administered 6 h prior to amygdala kindling markedly retards the development of amygdala-kindled after-discharges and seizures compared to animals either receiving ECS immediately after kindling or sham ECS [114]. These data indicate that ECS exerts marked anticonvulsive effects against limbic system seizures and raise the possibility that the limbic antiexcitability effects of electroconvulsive shock could potentially be related to its mechanism of clinical efficacy in affective illness. Thus, we suggest the possibility that the generalized seizures of electroconvulsive shock are themselves exerting profound anticonvulsive effects on limbic substrates.

A second potential explanation brings us back to a consideration of the controversial relationships of the seizure disorders to the development of psychopathology. Many investigators [17, 41, 42, 67–69, 80, 140] have noted the reciprocal relationship between seizures and psychosis. In fact, early observations of this relationship were instrumental in the initial introduction of electroconvulsive therapy into psychiatric practice. Thus, in some instances, seizures either endogenously or exogenously introduced, appear to exert positive effects on mood and behavioral disorders. Reciprocal observations have also been made. Investigators have noted the emergence of behavioral pathology following achievement of better anticonvulsant effects or in association with improvement in the EEG; i.e., the forced normalization of *Landolt* [68, 69]. These observations might help explain the therapeutic effects of electroconvulsive therapy on patients with affective disorders, but they would also predict an exacerbation of psychopathology following the introduction of anticonvulsant agents. Most of the reports of the reciprocal nature of seizures and psychosis have emphasized the occurrence of psychopathology in the schizophreniform spectrum. The exacerbation of schizophrenic symptomatology has been reported using several of the anticonvulsant compounds [70, 131]. We suggest the utility of the kindling [50, 51] model in systematically exploring in the laboratory these issues of the relationship between limbic seizures and behavior.

Implications of Kindling for the Development of Behavioral Pathology in Temporal Lobe Epilepsy and Alcohol Withdrawal Syndromes

We have observed a reciprocal relationship between seizures and behavioral pathology in some kindled seizure paradigms but a positive relationship in others. For example, in animals experiencing repeated amygdala-kindled seizures, there is a long-lasting deficit in spontaneous exploratory activity as well as a decreased responsiveness to cocaine-induced hyperactivity [111, 115, 117]. These findings have been replicated and extended by *Ehlers* et al. [34] who also demonstrated decreased responsivity to amphetamine following not only kindling of the amygdala but also kindling of the nucleus accumbens. In our studies, the decreased spontaneous and cocaine-induced motor activity following amygdala kindling was associated with an increased susceptibility to lidocaine-induced seizures [109, 117]. Thus, the amygdala kindling model may be useful for examining persisting behavioral alterations observed in the interictal period and possible reciprocal relationships between seizures and behavioral pathology. While the exact mechanisms underlying the decreasing motor responsivity following kindling remains to be determined, catecholamine and cholinergic changes have been reported [for reviews see 58, 78, 105, 117, 123, 132] and the altered responsivity following seizures provides an interesting model for examining how repeated limbic seizures may lead to changes in motor behavior. These findings may thus be of interest in relation to the partial overlap in symptoms of the interictal behavioral disturbance of temporal lobe epilepsy and manic-depressive illness with apparent exception of changes in the motor realm. These motor components might develop in the interictal period, possibly secondary to seizure-induced changes in catecholamine or cholinergic function.

Pharmacologically-induced kindling with repeated lidocaine administration provides a further means of studying the relationship of the development of behavioral pathology to seizures of the limbic system. Repeated subconvulsive doses of lidocaine are eventually associated with the production of tonic-clonic seizures resembling those achieved by electrophysiological amygdala kindling [109, 111, 115]. Following these seizures, animals show long-lasting persistence of aggressive responses as measured by resistance to capture [2, 109]. Our recent data suggest that this pharmacologically induced development of seizures and its associated behavioral changes are not related to alterations in lidocaine pharmacokinetics [*Post, Contel, Pitem,* unpubl. data, 1981]. Plasma levels of lidocaine are not higher in animals showing repeated lidocaine-induced seizures compared to lidocaine non-seizers or saline-pretreated controls acutely challenged with lidocaine. Moreover, blockade of lidocaine-induced seizures with diazepam blocks the development of the kindled seizure sensitization as well as the associated development of aggression.

Our studies in collaboration with *Kennedy* and *Sokoloff,* utilizing the ^{14}C-2-deoxyglucose autoradiographic method for measuring local glucose utilization, indicate that lidocaine seizures are producing increases in local metabolic activity in relatively discrete areas of the limbic system, including hippocampus and amygdala or perirhinal and cingulate cortical areas [109]. Activation of these limbic substrates appears to be the prerequisite for the development of aggressive behavior in animals, as this response does not occur in lidocaine-pretreated animals that do not show seizures or in animals pretreated with diazepam which blocks lidocaine seizures. Moreover, seizures induced by metrazol, which demonstrate different patterns of glucose utilization, or those induced by ECS, are not associated with the development of aggression.

Thus, the electrophysiological and pharmacological kindling paradigms may be of particular utility in the exploration of the direct and reciprocal relationships between seizures and the development of altered behavior. We have suggested that the behavioral consequences of kindling may be a useful conceptual approach to the late development of behavioral difficulties experienced by some patients with temporal lobe epilepsy [97, 98, 110]. In particular, the long lag between onset of seizures and the development of psychotic psychopathology (often 10–15 years) might be comprehensible from the kindling perspective. Not only does kindling increase duration, complexity, and spread of electrical activity, but this process itself or its indirect consequences on neurotransmitter function may eventually lead to the development of behavioral disturbances.

It is of interest that carbamazepine is useful in the treatment of the behavioral disorders not only associated with temporal lobe epilepsy and affective illness, but also those associated with alcohol withdrawal. Repeated episodes of alcohol withdrawal may themselves act as a repeated kindling stimulus [8]. A variety of behavioral and convulsive symptomatologies are associated with an increased duration of alcohol abuse and the associated increased number of episodes of alcohol withdrawal. Not only are seizures an eventual outcome of severe alcohol withdrawal reactions in a small percentage of patients, but disturbances of affect and cognition may develop in the course of severe alcohol intoxication and repeated withdrawal reactions. We have presented preliminary evidence elsewhere [8] that increasingly severe behavioral withdrawal symptomatology, including the possibility of DTs, may be associated with an increased duration and therefore number of alcohol withdrawal episodes. Preliminary data from both open and double-blind controlled studies suggest that carbamazepine may be of some use in the prevention of acute withdrawal episodes and their associated behavioral disturbances [16, 21, 23, 118]. These clinical data are consistent with those of *Chu* [24] indicating that carbamazepine is effective in blocking alcohol withdrawal reaction in mice.

We suggest the possibility that the interictal development of behavioral disturbances associated with repeated endogenous seizures of temporal lobe epilepsy and repeated episodes of neural excitation of alcohol withdrawal may, in part, be acting on common neural substrates underlying affective illness which are, in turn, responsive to the therapeutic effects of carbamazepine. It is also possible that the therapeutic effects of carbamazepine are unrelated to its temporal lobe-limbic antiexcitability properties; if this were the case, careful study of carbamazepine's physiological and biochemical effects should nevertheless prove rewarding. The kindling models appear to offer a unique perspective to the study of the progressive and long-lasting development of convulsive or behavioral alterations in several neuropsychiatric syndromes including temporal lobe epilepsy, alcohol withdrawal syndromes, and possibly, manic-depressive illness. This perspective has led us to studies of a new possible treatment of affective illness with carbamazepine and may be helpful in generating further conceptual and treatment approaches based on the use of anticonvulsants, peptides, and other agents which might interfere with the development of the long-term coding of pathological emotional behavior.

References

1. Adamec, R.: Behavioral and epileptic determinants of predatory attack behavior in the cat. Can. J. neurol. Sci. *2:* 457–466 (1975).
2. Albert, D.J.; Richmond, S.E.: Septal hyperreactivity. A comparison of lesions within and adjacent ot the septum. Physiol. Behav. *15:* 339–347 (1975).
3. Albright, P.S.; Burnham, W.M.: Development of a new pharmacological seizure model: effects of anticonvulsants on cortical- and amygdala-kindled seizures in the rat. Epilepsia *21:* 681–689 (1980).
4. Ashton, D.; Wauquier, A.: Behavioral analysis of the effects of 15 anticonvulsants in the amygdaloid kindled rat. Psychopharmacology *65:* 7–13 (1979).
5. Babington, R.G.: The pharmacology of kindling; in Hanin, Usdin, Animal models of psychiatry and neurology, pp. 141–149 (Pergamon Press, Oxford 1977).
6. Babington, R.G.; Horovitz, Z.P.: Neuropharmacology of SQ 10,996, a compound with several therapeutic indications. Archs. int. Pharmacodyn. Ther. *202:* 106–118 (1973).
7. Ballenger, J.C.; Post, R.M.: Therapeutic effects of carbamazepine in affective illness. A preliminary report. Commun. Psychopharmacol. *2:* 159–175 (1978).
8. Ballenger, J.C.; Post, R.M.: Kindling as a model for the alcohol withdrawal syndromes. Br. J. Psychiat. *133:* 1–14 (1978).
9. Ballenger, J.C.; Post, R.M.: Carbamazepine (Tegretol) in manic-depressive illness. A new treatment. Am. J. Psychiat. *137:* 782–790 (1980).
10. Bawden, H.N.; Racine, R.J.: Effects of bilateral kindling or bilateral subthreshold stimulation of the amygdala or septum on muricide, ranacide, intraspecific aggression and passive avoidance in the rat. Physiol. Behav. *22:* 115–123 (1979).

11 Bear, D.; Fedio, P.: Quantitative analysis of interictal behavior in temporal lobe epilepsy. Archs Neurol. *34:* 454–467 (1977).
12 Bernasconi, R.; Martin, P.: Effects of antileptic drugs on the GABA turnover rate. Arch. Pharmacol. suppl. 307. p. R63, abstr. 251 (1979).
13 Berrettini, W.H.; Hare, T.; Ballenger, J.C.; Post, R.M.: GABA metabolism in affective illness; in Post, Ballenger, Neurobiology of the mood disorders. (Williams & Wilkins, Baltimore, in press, 1981).
14 Berrettini, W.H.; Post, R.M.; Worthington, E.K.; Caspar, J.B.: Human platelet vasopressin receptors. Life Sci. (in press, 1982).
15 Bertilsson, L.: Clinical pharmacokinetics of carbamazepine. Clin. Pharmacokinet. *3:* 128–143 (1978).
16 Bjorkqvist, S.E.; Isohanni, M.; Makela, R.; Malinen, L.: Ambulant treatment of alcohol withdrawal symptoms with carbamazepine. A formal multicentre double-blind comparison with placebo. Acta psychiat. scand. *53:* 333–342 (1976).
17 Blumer, D.: Temporal lobe epilepsy and its psychiatric significance; in Benson, Blumer, Psychiatric aspects of neurologic disease, pp. 171–198 (Grune & Stratton, New York 1975).
18 Boast, C.A.; McIntrye, D.C.: Bilateral kindled amygdala foci and inhibitory avoidance behavior in rats. A functional lesion effect. Physiol. Behav. *18:* 25–28 (1977).
19 Bogoch, S.; Dreyfus, J.: The broad range of use of diphenylhydantoin. Bibliography and review, vol. I (The Dreyfus Medical Foundation, 1970).
20 Bogoch, S.; Dreyfus, J.: DPH, 1975. A supplemental to the broad range of use of diphenylhydantoin. Bibliography and review (The Dreyfus Medical Foundation 1975).
21 Brune, F.; Busch, H.: Anticonvulsive-sedative treatment of delirium alcoholism. Q.J. Stud. Alcohol. *32:* 334–342 (1971).
22 Bunney, W.E., Jr.; Hamburg, D.A.: Method for reliable longitudinal observation of behavior. Archs gen. Psychiat. *9:* 280–294 (1963).
23 Carlsson, C.; Pettersson, L.: Dysphoric symptoms in chronic alcoholics and the effects of carbamazepine (Tegretol). Int. J. clin. Pharmacol. *5:* 403–405 (1972).
24 Chu, N.-S.: Carbamazepine. Prevention of alcohol withdrawal seizures. Neurology, Minneap. *29:* 1397–1401 (1979).
25 Cutler, N.R.; Post, R.M.: Life course of untreated manic-depressive illness. Sci. Proc. Am. Psychiat. Ass. *133:* 71, abstr. 28 (1980).
26 Dalby, M.A.: Antiepileptic and psychotropic effect of carbamazepine (Tegretol) in the treatment of psychomotor epilepsy. Epilepsia *12:* 325–334 (1971).
27 Dalby, M.A.: Behavioral effects of carbamazepine; in Penry, Daly, Complex partial seizures. Advances in neurology, vol. 11, pp. 331–343 (Raven Press, New York 1975).
28 Deflorida, F.A.; Delgado, J.M.R.: Lasting behavioral and EEG changes in cats induced by prolonged stimulation of amygdala. Am. J. Physiol. *193:* 223–229 (1958).
29 Delgado, J.M.R.; Anand, B.K.: Increase of food intake induced by electrical stimulation of the lateral hypothalamus. Am. J.Physiol. *172:* 162–168 (1953).
30 Delgado, J.M.R.; DeFeudis, F.V.: Effects of lithium injections into the amygdala and hippocampus of awake monkeys. Expl Neurol. *25:* 255–267 (1967).
31 de Wied, D.: Peptides and behavior. Life Sci. *20:* 195–204 (1977).
32 de Wied, D.: Peptides and adaptive behavior; in de Wied, van Keep, Hormones and behavior, pp. 103–113 (MTP Press, Lancaster 1980).

33 Dongier, S.: Statistical study of clinical and EEG manifestations of 536 psychotic episodes occurring in 516 epileptics between clinical seizures. Epilepsia *1:* 117–142 (1959).
34 Ehlers, C.L.; Koob, G.F.; Henriksen, S.J.; Bloom, F.E.: Changes in locomotor activity in rats 'kindled' in the amygdala and nucleus accumbens septi. Abstr. Soc. Neurosci. *6:* 399, abstr. 137.2 (1980).
35 Emrich. H.M.; Zerssen, D.v.; Altmann, H.; Kissling, W.; Möller, H.-J.: The effect of GABA-ergic drugs in mania. Abstr. 3rd Wld Congr. Biological Psychiatry, Stockholm 1971, Part I, abstr. S 304.
36 Emrich, H.M.; Zerssen, D.v.; Kissling, W.; Möller, H.-J.; Windorfer, A.: Effect of sodium valproate in mania. The GABA-hypothesis of affective disorders. Arch. Psychiat. NervKrank. *229:* 1–16 (1980).
37 Ervin, F.R.; Delgado, J.; Mark, V.H.; Sweet, W.H.: Rage. A paraepileptic phenomenon? Epilepsia *10:* 417 (1969).
38 Faigle, J.W.; Feldmann, K.F.; Baltzer, V.: Anticonvulsant effect of carbamazepine. An attempt to distinguish between the potency of the parent drug and its epoxide metabolite; in Gardner-Thorpe, Janz, Meinardi, et al., Antiepileptic drug monitoring, pp. 104–109 (Pitman Press, Avon 1977).
39 Fink, M.: Convulsive therapy. Theory and practice (Raven Press, New York 1979).
40 Frigerio, A.; Morselli, P.L.: Carbamazepine. Biotransformation; in Penry, Daly, Complex Partial seizures. Advances in neurology, vol. 11, pp. 295–308 (Raven Press, New York 1975).
41 Flor-Henry, P: Psychosis and temporal lobe epilepsy. A controlled investigation. Epilepsia *10:* 363–395 (1969).
42 Flor-Henry, P.: Schizophrenic-like reactions and affective psychoses associated with temporal lobe epilepsy. Etiological factors. Am. J. Psychiat. *126:* 400–403 (1969).
43 Fromm, G.H.; Chattha, A.S.; Terrence, C.F.; Glass, J.D.: Role of inhibitory mechanisms in trigeminal neuralgia. Neurology, Minneap. (in press, 1981).
44 Fromm, G.H.; Glass, J.D.; Chattha, A.S.; Martinez, A.J.: Effect of anticonvulsant drugs on inhibitory and excitatory pathways. Epilepsia *22:* 65–73 (1981).
45 Gagneux, A.R.: The chemistry of carbamazepine; in Birkmayer, Epileptic seizures-behavior-pain, pp. 120–126 (Huber, Bern 1976).
46 Ghose, K.: Effect of carbamazepine in polyuria associated with lithium therapy. Pharmakopsychiatria *11:* 241–245 (1978).
47 Gibbs, F.A.: Ictal and nonictal psychiatric disorders in temporal lobe epilepsy. J.nerv. ment. Dis. *113:* 522–528 (1951).
48 Glaser, G.H.; Newman, R.J.; Schafter, R.: Interictal psychosis in psychomotor-temporal lobe epilepsy. An EEG-psychologial study; in Glaser, EEG and behavior, pp. 345–365 (Basic Books, new York 1963).
49 Gloor, P.: Temporal lobe epilepsy. Its possible contribution to the understanding of the functional significance of the amygdala and of its interaction with neocortical-temporal mechanisms; in Eleftheriou, The neurobiology of the amygdala, pp. 423–457 (Plenum Press, New York 1972).
50 Goddard, G.V.; McIntyre, D.C.; Leech, C.K.: A permanent change in brain function resulting from daily electrical stimulation. Expl Neurol. *25:* 295–330 (1969).
51 Goddard, G.V.; Morrell, F.: Chronic progressive epileptogenesis induced by focal electrical stimulation of brain. Neurology, Minneap. *21:* 393 (1971).
52 Gold, P.W.; Goodwin, F.K.; Ballenger, J.C.; Weingartner, H.; Robertson G.L.; Post,

R.M.: Central vasopressin function in affective illness; in de Wied, van Keep, Hormones and the brain, pp. 241–252, (MTP Press, Lancaster 1980).

53 Gold, P.W.; Reus, V.I.; Goodwin, F.K.: Vasopressin in affective illness – hypothesis. Lancet *i:* 1233–1235 (1978).

54 Goodwin, F.K.; Murphy, D.L.; Bunney, W.E., Jr.: Lithium carbonate treatment in depression and mania: a longitudinal double-blind study. Archs gen. Psychiat. *21:* 486–496 (1969).

55 Goodwin, F.K.; Post, R.M.; Dunner, D.L.; Gordon, E.K.: Cerebrospinal fluid amine metabolites in affective illness. The probenecid technique. Am. J. Psychiat. *130:* 73–79 (1973).

56 Grof, P.; Angst, J.; Haines, T.: The clinical course of depression. Practical issues; in Schattauer, Symposia medica Hoest, vol. 8: Classification and prediction of outcome of depression, pp. 141–148 (Schattauer, New York 1974).

57 Grof, P.; Zis, A.P.; Goodwin, F.K.; Wehr, T.A.: Patterns of recurrence in bipolar affective illness. Sci. Proc. Am. Psychiat. Ass. *131:* 179–180, abstr. 349 (1978).

58 Gunne, L.M.; Reis, D.J.: Changes in brain catecholamines associated with electrical stimulation of amygdaloid nucleus. Life Sci. *2:* 804–809 (1963).

59 Heath, R.G.: Correlation of brain function with emotional behavior. Biol. Psychiat. *11:* 463–480 (1976).

60 Heath, R.G.; Monroe, R.R.; Mickle, W.: Stimulation of the amygdaloid nucleus in a schizophrenic patient. Am. J. Psychiat. *111:* 862–863 (1955).

61 Henriksen, S.J.; Bloom, F.E.; McCoy, F.; Ling, N.; Guillemin, R.: Beta-endorphin induces nonconvulsive limbic seizures. Proc. natn. Acad. Sci. USA *75:* 5221–5225 (1978).

62 Innes, E.B.; Locatell-Innes, E.M.; Cornell, J.M.: Amygdaloid kindled convulsions and modification of weight gains in hooded rats. Physiol. Behav. *18:* 187–192 (1977).

63 Kalinowsky, L.B.; Putnam, T.J.: Attempts at treatment of schizophrenic and other nonepileptic psychoses with dilantin. Arch Neurol. Psychiat. *49:* 414–420 (1943).

64 Katz, R.J.; Schmaltz, K.: Facilitation of opiate- and enkephalin-induced motor activity in the mouse by phenytoin sodium and carbamazepine. Psychopharmacology *65:* 65–68 (1979).

65 Kraepelin, E.: Manic-depressive insanity and paranoia (Livingstone, Edinburgh 1920).

66 Lambert, P.A.; Carraz, G.; Borselli, S.; Bouchardy, M.: Le dipropylacétamide dans le traitement de la psychose maniaco-dépressive. Encéphale *1:* 25–31 (1975).

67 Lamprecht, F.: Epilepsy and schizophrenia. A neurochemical bridge. J. neural. Transm. *40:* 159–170 (1977).

68 Landolt, H.: Some clinical electroencephalographic correlations in epileptic psychoses. Electroenceph. Clin. Neurophysiol. *5:* 121 (1953).

69 Landolt, H.: Serial EEG investigations during psychotic episodes in epileptic patients and during schizophrenic attacks; in Lorentz de Haas, Lectures on epilepsy, pp. 91–133 (Elsevier, Amsterdam 1958).

70 Lautin, A.; Angrist, B.; Stanley, M.; Gershon, S.; Heckl, K.; Karobath, M.: Sodium valproate in schizophrenia. Some biochemical correlates. Br. J. Psychiat. *137:* 240–244 (1980).

71 Le Gal La Salle, G.: Inhibition of kindling-induced generalized seizures by aminooxyacetic acid. Can. J. Physiol. Pharmacol. *58:* 7–11 (1979).

72 Leviel, V.; Naquet, R.: A study of the action of valproic acid on the kindling effect. Epilepsia *18:* 229–234 (1977).

73 MacLean, P.D.: Some psychiatric implications of physiological studies on frontotemporal portion of limbic system (visceral brain). Electroenceph. clin. Neurophysiol. *4:* 407–418 (1952).
74 MacLean, P.D.: Chemical and electrical stimulation of hippocampus in unrestrained animals. II. Behavioral findings. Arch Neurol. Psychiat. *78:* 128–142 (1957).
75 MacLean, P.D.: The limbic brain in relation to the psychoses; in Black, Physiological correlates of emotion (Academic Press, New York 1970).
76 MacLean, P.D.: A triune concept of the brain and behavior; in Boag, Campbell, The Clarence M. Hicks Memorial Lectures, 1969 (University of Toronto Press, Toronto 1973).
77 Martinon, J.M.; Docampo, G.; Martinon, F.; Viso, J.A.; Pena, J.: Distonia por carbamacepina: a propósito de cuatro observaciones en el niño. An. exp. Pediat. *13:* 789–792 (1980).
78 McIntyre, D.C.: Kindling and memory: the adrenal system and the bisected brain, in Livingston, Hornykiewicz, Limbic mechanisms. The continuing evolution of the limbic system concept, pp. 495–506 (Plenum Press, New York 1978).
79 McIntyre, D.C.; Molino, A.: Amygdala lesions and CER learning – long term effects of kindling. Physiol. Behav. *8:* 1055–1058 (1972).
80 Monroe, R.: Episodic behavioral disorders (Harvard University Press, Cambridge 1970).
81 Murphy, D.L.; Kalin, N.: Utilization of substrate selective monoamine oxidase inhibitors which are antidepressants to explore neurotransmitter hypotheses of the affective disorders; in Post, Ballenger, Neurobiology of the mood disorders (Williams & Wilkins, Baltimore, in press, 1981).
82 Murphy, D.L.; Pickar, D.; Alterman, I.S.: Methods for the quantitative assessment of depressive and manic behavior; in Burdock, Sudilovsky, Gershon, Quantitative techniques for the evaluation of the behavior of psychiatric patients (Marcel Dekker, New York, in press, 1981).
83 Myslobodsky, M.S.; Ackermann, R.F.; Engel, J.: Effects of γ-acetylenic GABA and γ-vinyl GABA on metrazol-activated and kindled seizures. Pharmacol. Biochem. Behav. *11:* 265–271 (1979).
84 Myslobodsky, M.S.; Valenstein, E.S.: Amygdaloid kindling and the GABA system. Epilepsia *21:* 163–175 (1980).
85 Okuma, T.; Inanaga, K.; Otsuki, S.; Sarai, K.; Takahashi, R.; Hazama, H.; Mori, A.; Watanabe, M.: Comparison of the antimanic efficacy of carbamazepine and chlorpromazine. A double-blind controlled study. Psychopharmacology *66:* 211–217 (1979).
86 Okuma, T.; Inanaga, K.; Otsuki, S.; Sarai, K.; Takashashi, R.; Hazama, H.; Mori, A.; Watanabe, M.: A preliminary double-blind study of the efficacy of carbamazepine in prophylaxis of manic-depressive illness. Psychopharmacology *73:* 95–96 (1981).
87 Okuma, T.; Kishimoto, A.; Inoue, K.; Matsumoto, H.; Ogura, A.; Matsushita, T.; Naklao, T.; Ogura, C.: Anti-manic and prophylactic effects of carbamazepine on manic-depressive psychosis. Folia psychiat. neurol. jap. *27:* 283–297 (1973).
88 Overall, J.E.; Gorham, D.R.: The brief psychiatric rating scale. Psychiat. Rep. *10:* 799–812 (1962).
89 Papez, J.W.: A proposed mechanism of emotion. Archs Neurol. Psychiat. *38:* 725–743 (1937).
90 Penry, J.K.; Daly, D.D.: Complex partial seizures. Advances in neurology, vol. 11 (Raven Press, New York 1975).

91 Perucca, E.; Garratt, A.; Hebdige, S.; Richens, A.: Water intoxication in epileptic patients receiving carbamazepine. J. Neurol. Neurosurg. Psychiat. *41:* 713–718 (1978).
92 Pinel, J.P.J.; Rovner, L.I.: Experimental epileptogenesis. Kindling-induced epilepsy in rats. Expl Neurol. *58:* 190–202 (1978).
93 Pinel, J.P.J.; Rovner, L.I.: Electrode placement and kindling-induced experimental epilepsy. Expl Neurol. *58:* 335–346 (1978).
94 Pinel, J.P.J.; Treit, D.; Rovner, L.I.: Temporal lobe aggression in rats. Science *197:* 1088–1089 (1977).
95 Pisciotta, A.V.: Hematologic toxicity of carbamazepine; in Penry, Daly, Complex partial seizures. Advances in neurology, vol. 11, pp. 355–366 (Raven Press, New York 1975).
96 Post, R.M.: Cocaine psychosis. A continuum model. Am. J. Psychiat. *132:* 225–231 (1975).
97 Post, R.M.: Clinical implications of a cocaine-kindling model of psychosis, in Klawans, Clinical neuropharmacology, vol. II, pp. 25–42 (Raven Press, New York 1977).
98 Post, R.M.; Ballenger, J.C.: Kindling models for the progressive development of behavioral psychopathology. Sensitization to electrical, pharmacological, and psychological stimuli; in van Praag, Lader, Rafaelsen, Sachar, Handbook of Biological psychiatry, part IV, pp. 609–651 (Marcel-Dekker, New York 1981).
99 Post, R.M.; Ballenger, J.C.; Hare, T.A.; Bunney, W.E., Jr.: Lack of effect of carbamazepine on gamma-aminobutyric acid levels in cerebrospinal fluid. Neurology, Minneap. *30:* 1008–1011 (1980).
100 Post, R.M.; Ballenger, J.C.; Hare, T.A.; Goodwin, F.K.; Lake, C.R.; Jimerson, D.C.; Bunney, W.E., Jr.: Cerebrospinal fluid GABA in normals and patients with affective disorders. Brain Res. Bull. *5:* 755–759 (1980).
101 Post, R.M.; Ballenger, J.C.; Reus, V.I.; Lake, C.R.; Lerner, P.; Bunney, W.E., Jr.: Effects of carbamazepine in mania and depression. 131st Ann. Meet. Am. Psychiat. Ass. Atlanta 1978, new research abstr. 7.
102 Post, R.M.; Ballenger, J.C.; Rey, A.C.; Bunney, W.E., Jr.: Slow and rapid onset of manic episodes: implications for underlying biology. Psychiat. Res. *4:* 229–237 (1981).
103 Post, R.M.; Ballenger, J.C.; Uhde, T.W.; Bunney, W.E., Jr.: Carbamazepine in manic-depressive illness: implications for underlying mechanisms; in Post, Ballenger, Neurobiology of the mood disorders (Williams & Wilkins, Baltimore, in press, 1982).
104 Post, R.M.; Ballenger, J.C.; Uhde, T.W.; Chatterji, D.C.; Bunney, W.E., Jr.: Efficacy of the temporal lobe-limbic anticonvulsant carbamazepine in manic and depressive illness; in Jansson, Perris, Struwe, Proc. 3rd Wld Congr. Biological Psychiatry, Stockholm 1981 (Elsevier, Amsterdam, in press, 1981).
105 Post, R.M.; Ballenger, J.C.; Uhde, T.W.; Putnam, F.W.; Bunney, W.E., Jr.: Kindling and drug sensitization. Implications for the progressive development of psychopathology and treatment with carbamazepine; in The psychopharmacology of anticonvulsants. The British Association for Pharmacology Monograph Series (Oxford University Press, Oxford, in press, 1981).
106 Post, R.M.; Ballenger, J.C.; Uhde, T.W.; Smith, C.; Rubinow, D.R.; Bunney, W.E., Jr.: Effect of carbamazepine on cyclic nucleotides in CSF of patients with affective illness. Biol. Psychiat. (in press, 1982).
107 Post, R.M.; Gerner, R.H.; Carman, J.S.; Gillin, J.C.; Jimerson, D.C.; Goodwin, F.K.; Bunney, W.E., Jr.: Effects of a dopamine agonist piribedil in depressed patients. Rela-

tionship of pretreatment HVA to antidepressant response. Archs gen. Psychiat. *35:* 609–615 (1978).
108 Post, R.M.; Jimerson, D.C.; Bunney, W.E., Jr.; Goodwin, F.K.: Dopamine and mania: behavioral and biochemical effects of the dopamine receptor blocker pimozide. Psychopharmacology *67:* 297–305 (1980).
109 Post, R.M.; Kennedy, C.; Shinohara, M.; Squillace, K.M.; Miyoaka, M.; Suda, S.; Ingvar, D.H.; Sokoloff, L.: Local cerebral glucose utilization in lidocaine-kindled seizures. Abstr. 9th Ann. Meet. Neuroscience, vol. *5:* p. 196, abstr. 646 (1979).
110 Post, R.M.; Kopanda, R.T.: Cocaine, kindling, and psychosis. Am. J. Psychiat. *133:* 627–634 (1976).
111 Post, R.M.; Kopanda, R.T.; Lee, A.: Progressive behavioral changes during chronic lidocaine administration. Relationship to kindling. Life Sci. *17:* 943–950 (1975).
112 Post, R.M.; Lake, C.R.; Jimerson, D.C.; Bunney, W.E., Jr.; Wood, J.H.; Ziegler, M.G.; Goodwin, F.K.: Cerebrospinal fluid norepinephrine in affective illness. Am. J. Psychiat. *135:* 907–917 (1978).
113 Post, R.M.; Pickar, D.; Naber, D.; Ballenger, J.C.; Uhde, T.W.; Bunney, W.E., Jr.: Effect of carbamazepone on CSF opiod activity. Relationship to antidepressant response. Psychiat. Res. *5:* 59–66 (1981).
114 Post, R.M.; Putnam, F.W.; Contel, N.R.: Electroconvulsive shock inhibits amygdala kindling. Ann. Meet. Soc. Neuroscience, Los Angeles 1981.
115 Post, R.M.; Squillace, K.M.; Pert, A.: Kindling oscillations in amygdala excitability. Abstr. Soc. Biol. Psychiat. p. 38, abstr. 27 (1977).
116 Post, R.M.; Squillace, K.M.; Pert, A.: Rhythmic oscillations in amygdala excitability during kindling. Life Sci. *22:* 717–726 (1978).
117 Post, R.M.; Squillace, K.M.; Pert, A.; Sass, W.: The effects of amygdala kindling on spontaneous and cocaine-induced motor activity and lidocaine seizures. Psychopharmacology *72:* 189–196 (1981).
118 Poutanen, P.: Experience with carbamazepine in the treatment of withdrawal symptoms in alcohol abusers. Br. J. Addict. *74:* 201–204 (1979).
119 Purdy, R.E.; Julien R.M.; Fairhurst, A.S.; Terry, M.D.: Effect of carbamazepine on the in vitro uptake and release of norepinephrine in adrenergic nerves of rabbit aorta and in whole brain synaptosomes. Epilepsia *18:* 251–257 (1977).
120 Putnam, F.W.; Contel, N.R.; Post, R.M.: Amygdala kindling in vasopressin deficient rats (unpubl. data, 1981).
121 Quattrone, A.; Crunelli, V.; Samanin, R.: Seizure susceptibility and anticonvulsant activity of carbamazepine, diphenylhydantoin and phenobarbital in rats with selective depletions of brain monoamines. Neuropharmacology *17:* 643–647 (1978).
122 Quattrone, A.; Samanin, R.: Decreased anticonvulsant activity of carbamazepine in 6-hydroxydopamine-treated rats. Eur. J. Pharmacol. *41:* 333–336 (1977).
123 Racine, R.J.: Kindling. The first decade. Neurosurgery *3:* 234–252 (1978).
124 Reus, V.I.; Targum, S.D.; Weingartner, H.; Post, R.M.: Effects of lithium carbonate on memory processes of bipolar affectively ill patients. Psychopharmacology *63:* 39–42 (1979).
125 Rubinow, D.R.; Gold, P.W.; Ballenger, J.C.; Post, R.M.; Goodwin, F.K.: Somatostatin; in Post, Ballenger, Neurobiology of the mood disorders (Williams & Wilkins, Baltimore, in press, 1981).

126 Sherwin, I.: Psychosis associated with epilepsy: significance of the laterality of the epileptogenic lesion. J. Neurol. Neurosurg. Psychiat. *44:* 83–85 (1981).
127 Silberman, E.K.; Post, R.M.: The 'march' of symptoms in a psychotic decompensation. Case report and theoretical implications. J. nerv. ment. dis. *168:* 104–110 (1980).
128 Spitzer, R.L.; Endicott, J.; Robins, E.: Research diagnostic criteria. Rationale and reliability. Archs gen. Psychiat. *35:* 773–782 (1978).
129 Stephens, W.P.; Coe, J.Y.; Baylis, P.H.: Plasma arginine vasopression concentrations and antidiuretic action of carbamazepine. Br. med. J. *i:* 1445–1447 (1978).
130 Stevens, J.R.: Interictal clinical manifestations of complex partial seizures; in Penry, Daly, partial seizures. Advances in neurology, vol. 11. pp. 85–112 (Raven Press, New York 1975).
131 Stevens, J.R.; Bigelow, L.; Denney, D.; Lipkin, J.; Livermore, A.; Ranscher, F.; Wyatt, R.J.: Telemetered EEG-EOG during psychotic behaviors of schizophrenia. Archs gen. Psychiat. *36:* 251–262 (1979).
132 Stevens, J.R.; Livermore, A., Jr.: Kindling of the mesolimbic dopamine systems. Animal model of psychosis. Neurology, Minneap. *28:* 36–46 (1978).
133 Stevens, J.R.; Mark, V.H.; Erwin, F.; Pacheco, P.; Suematsu, K.: Deep temporal stimulation in man: long latency, long lasting psychological changes. Archs Neurol. *21:* 157–167 (1969).
134 Stokes, K.A.; McIntyre, D.C.: Lateralized asymmetrical state-dependent learning produced by kindled convulsions from rat hippocampus. Physiol. Behav. *26:* 163–169 (1981).
135 Takezaki, H.; Hanaoka, M.: The use of carbamazepine (Tegretol) in the control of manic-depressive psychosis and other manic, depressive states. Clin. Psychiat. *13:* 173–183 (1971).
136 Tanaka, T.; Lange, H.: L'effet d'embrasement (kindling effect) par stimulation amygdadienne chez le chat et le rat: approche neurophysiologique et neuropharmacologique. Revue Eléctroenceph. Neurophysiol. Clin. *5:* 41–44 (1975).
137 Taylor, D.C.: Factors influencing the occurrence of schizophrenia-like psychosis in patients with temporal lobe epilepsy. Psychol. Med. *5:* 249–254 (1975).
138 Thomas, T.H.; Ball, S.G.; Wales, J.K.; Lee, M.R.: Effect of carbamazepine on plasma and urine arginine-vasopressin. Clin. Sci. mol. Med. *54:* 419–424 (1978).
139 Thompson, P.; Huppert, F.; Trimble, M.: Anticonvulsant drugs, cognitive function and memory. Acta neurol. scand. *62:* suppl. 80, pp. 75–81 (1980).
140 Trimble, M.: The relationship between epilepsy and schizophrenia. A biochemical hypothesis. Biol. Psychiat. *12:* 299–304 (1977).
141 Turner, W.J.: The usefulness of diphenylhydantoin in treatment of non-epileptic emotional disorders. Int. J. Neuropsychiat. *3:* suppl. 2 pp. S 8–S 20 (1967).
142 Uhde, T.W.; Post, R.M.: Hyponatremia during carbamazepine administration (unpubl. data 1981).
143 Valenstein, E. S.; Cox, V.C.; Kakolewski, J.W.: Reexamination of the role of the hypothalamus in motivation. Psychol. Rev. *77:* 16–31 (1970).
144 Wada, J.A.: Pharmacological prophylaxis in the kindling model of epilepsy. Archs. Neurol. *34:* 389–395 (1977).
145 Wada, J.A.; Sato,M.: Generalized convulsive seizures induced by daily electrical stimulation of the amygdala in cats. Correlative electrographic and behavioral features. Neurology, Minneap. *24:* 565–574 (1974).

146 Wada, J.A.; Sato, M.; Corcoran, M.E.: Persistant seizure susceptibility and recurrent spontaneous seizures in kindled cats. Epilepsia *15:* 465–478 (1974).
147 Wada, J.A.; Sato, M.; Wake, A.; Green, J.R.; Troupin, A.S.: Prophylactic effects of phenytoin, phenobarbital, and carbamazepine examined in kindled cat preparations. Archs Neurol. *33:* 426–434 (1976).
148 Wales, J.K.: Treatment of diabetes insipidus with carbamazepine. Lancet *ii:* 948–951 (1975).
149 Weingartner, H.; Gold, P.W.; Ballenger, J.C.; Smallberg, S.A.; Summers, R.; Rubinow, D.R.; Post, R.M.; Goodwin, F.K.: Effects of vasopressin on human memory functions. Science *211:* 601–603 (1981).
150 Weingartner, H.; White, N.: Exploration evoked by electrical stimulation of amygdala of rats. Physiol. Psychol. *6:* 229–235 (1978).
151 Wise, R.A.: Hypothalamic motivational systems. Fixed or plastic neural circuits. Science *162:* 377–379 (1968).
152 Yoshii, N.; Yamaguchi, Y.: Conditioning of seizure discharges with electrical stimulation of the limbic structures in cats. Folia psychiat. neurol. jap. *17:* 269–286 (1963.

R.M. Post, MD, National Institute of Health, 9000 Rockville Pike, Building 10, Room 3S239, Bethesda, MD 20205 (USA)

Epilogue

Werner P. Koella, Michael R. Trimble

Friedrich Miescher Institute, Basel, Switzerland; Institute of Neurology, London, England

This symposium was initiated and organized with a number of statements, facts and also questions in mind. It seems that some of these statements could be substantiated, while others turned out to be less clear. To some of our questions we obtained answers; to others we have as yet no good reply. Some of the questions possibly turned out to be inadequate in the sense that they remain, and may forever remain, unanswerable.

In viewing the way that man has looked at mental illness in the past we are aware of several dominant historical phases. *Hippocrates,* in the 4th century BC, could postulate: 'that from nothing else but thence, from the brain, comes joys, delights, laughter and sports, and sorrows, griefs, despondency, and lamentations ... And by this same organ we become mad and delirious, and fears and terrors assail us ...' [1]. Still, such a farsighted statement was soon overlooked and forgotten. Behavior, especially deviant behavior, became to be seen as driven by demons and gods, and indeed a subtle distinction was made between epilepsy, the sacred disease, and mania or madness, on the grounds that patients with the latter were possessed by demons, whereas only a god could throw a sane man to the ground, deprive him of his senses, and then restore him to normality [4].

Following the renaissance, and especially in the 19th century, great progress was made towards improving the understanding of the nervous system; yet, a failure to identify lesions in the brains of many patients with insanity led to a divergence of opinion regarding the etiology and pathogenesis of neurological disease, in which lesions could be found and demonstrated, and psychiatric illness – the latter eventually being seen as the result of psychological conflict and trauma, not now from some external malevolent force, but from internal drives, some of which were equally conceptually evil.

It is only in the midpoint of this century, however, that a re-statement of Hippocratic ideas has been clearly based on a scientific foundation with the acknowledgement that, however substantial ideas of the id and ego turn out to be, such conceptualized structures require neuronal substrates for their understanding, and that through further exploration of the central nervous system we will eventually come to understand our own behavior. The evolution of the limbic system concept, especially through the writings of *Papez* [7], *Yakovlev* [12] and *MacLean* [5], has occurred pari passu with the accumulation of data regarding the intimate relationships of the limbic system to behaviour both in animals and in man, as reviewed by *Koella*. While it may seem, on account of the vast growth of the neocortex in man, that the limbic system of homo sapiens has shrunk to a size of unimportance, we know now that this is not the case. The hippocampus, septum, entorhinal area and related limbic system are actually four to five times larger in man than in insectivores, on an equivalent body weight basis, and in some regions for example the CA_1 of the hippocampus the increase in size is even greater [9]. While there are differences between the behaviors of lower animals and man which follow stimulation or lesion within the limbic system, there are also similarities, so much so that we have begun to look to the limbic system as a possible site of abnormality in patients who display disturbed behavior, and to describe the exact patterns of behavior of patients with limbic system lesions for clues, links, and mechanisms of our understanding of the neural basis of behavior.

In our introduction we made a statement that concerned the symptomatological overlap between temporal lobe seizures – psychomotor fits – psychosis, in particular acute schizophreniform episodes, and florid manic attacks. *Stevens* [10], like many before and after her, pointed towards the surprising 'sharing of numerous symptoms between schizophreniform and temporal lobe epileptic attacks'. The data accumulated and presented here by *Sherwin* and *Trimble and Perez* add credence to the notion that an interictal schizophrenia-like psychosis occurs in epilepsy, in particular in relation to dominant side temporal lobe lesions. This psychosis associated with epilepsy is shown to be truly 'schizophrenia-like', and is 'on balance phenomenologically more like than different from process schizophrenia.' In addition, *Post* et al. draw similarities between manic-depressive symptoms and those seen in temporal lobe epilepsy, although further data on this issue are desirable before firm conclusions can be reached.

It is nevertheless clear that not all patients with temporal lobe epilepsy develop psychopathology, and *Stevens,* looking at the whole issue in a more 'statistical' manner has noted certain risk factors that in patients with epilepsy, and more specifically in temporal lobe epilepsy, may predispose to a high incidence of psychiatric disturbance. While allowing for a higher risk in such states, she cautions against too high estimates, which may be due, for example, to improper sampling techniques. For *Stevens* such risk factors include evidence for 'deep' or 'diffuse structural abnormalities' of the brain, basal or multiple spike foci, automatisms or visceral auras, and also a postchildhood onset of seizures and abnormal findings on neurological examination. With regard to risk factors specifically for aggressive behavior she singles out a low IQ, poor socioeconomic environment, early age of onset of seizures, and akinetic rather than pure temporal lobe episodes.

If we accept such data we may then vouch for a major – necessary yet not sufficient – pathophysiological condition in the temporal lobe of patients with epilepsy and psychosis. Electroencephalographic findings, though not necessarily conclusive, strongly suggest that pathology occurs in this area in many cases of psychomotor epilepsy and the anatomical work of *Scheibel* et al. [8] and others showing grave structural alterations of the neuronal and glial elements in the temporal lobe of patients with complex-partial seizures further supports this notion.

However, to date the evidence relating limbic, or more specifically temporal, lesions to the pathogenesis of psychosis is considerably less straightforward. *Stevens* has demonstrated and discussed focal pathology, including hamartomas, horizontal orientation of the hippocampal pyramidal cells, and vascular abnormalities, which occur in the temporal lobes of patients with epilepsy and psychosis, some of which changes may also be seen in schizophrenia. In addition, however, pathology elsewhere, such as periventricular fibrillary gliosis is noted, raising to question the possibility of 'downstream' changes that may be responsible for the psychotic development – a theme more fully developed elsewhere [10, 11]. *Stevens* in particular has suggested the frontobasal limbic area, especially the dopamine-receptive nucleus accumbens, as playing a strategic role in the generation of temporal lobe epileptic symptomatology *and* psychosis. In noting that catecholaminergic-blocking neuroleptic drugs not only are valuable to combat violence, paranoia and hallucinations, but also increase the propensity for epileptic seizures, she has suggested that the catecholamines may be the brain's natural defence against propagation of focal epileptic discharges. It is difficult to ignore the major role now ascribed to the catecholamines in the pathogenesis of psychiatric disorders, in particular dopamine in schizophrenia. Further exploration of the ways by which local synaptic mechanisms, in particular 'wet' neurotransmitters, can organize and disorganize behavior, is required, and some work on the essential role of the hippocampus in this respect has been presented by *Anderson.*

Animal experiments – although such evidence must always be looked at with important species-specific differences in mind – suggest that a hippocampal 'lesion' [in *MacLean's* experiments a local seizure, see *Koella's* paper] produces in the cat a behavior that can easily be interpreted – mutatis mutandis – as being 'psychotic'. One should also be aware that LSD-25 (still one of our best psychotomimetic agent) acts at least in part by suppression of the serotonergic neuron discharge in the raphe nuclei and thus via reduced release in the various serotonergic projection areas, viz. the hippocampus and other limbic structures [2]. Also LSD produces and enhances, in particular in the hippocampus, the typical theta-waves which we are bound to encounter too during the 'physiological hallucinatory' state, namely paradoxical or REM sleep.

Finally, and in addition to the aforementioned observations of *MacLean,* the many psychophysiological animal experiments seem to indicate that manipulations in the various limbic structures produce comportmental changes very much reminiscent – mutatis mutandis – of abnormal, if not psychotic behavior. In addition to schizophrenia we have also mentioned that symptoms of mania tend to overlap with both schizophrenia and temporal lobe seizures. Again, an in part common pathophysiology, both in locus and nature, may be envisioned as the ultimate underlying cause for this resemblence. Unfortunately, little is yet known about the structural and functional pathology of mania, with the possible exception of some ideas regarding exaggerated noradrenergic activity as presented in the well-established 'switch' theory [3]. However, *Okuma* et al. [6] and *Post* [this volume] have presented excellent evidence that carbamazepine, the drug of choice for the treatment of temporal lobe epilepsy, is quite efficient in combating mania and possibly in preventing manic-depressive attacks. This overlap in therapeutic efficiency may well be taken as additional evidence that the two afflictions – temporal lobe epilepsy and mania, if not also depression – may again be the manifestations of an in-part common pathology.

In turn there is 'therapeutic overlap' between mania and schizophrenia, neuroleptics being used to combat both. One theory of schizophrenia is related to overactivity of dopaminergic systems, particularly in the mesolimbic and/or mesohippocampal and amygdala systems. While the dopamine receptor blocking activity of the neuroleptics appears to be of utmost importance for the ability of these drugs to treat schizophrenia, in mania there is yet no such clue as to the reason for their therapeutic efficiency, and further evidence on this point remains to be shown by further work.

In spite of the vast gaps in our knowledge at the present time, we would like to speculate that all three of the afflictions that we have been discussing, namely schizophrenia, mania and temporal lobe epilepsy, do derive in part – spatially and in nature – from a common pathology, the overlapping symptomatology being *one* manifestation of this. Of course we recognize much nonoverlapping symptomatology, which is related to additional pathophysiological events. Nevertheless, if we are further to understand the subtle interrelationships between exteroceptive (environmental) and interoceptive influences on our behavior, we believe that the false dichotomy that has arisen in the neurosciences between psychiatric and neurological illness, the former supposedly wholly environmentally determined (the brainless mind) and the latter entirely endogenous (the mindless brain), needs to be overcome. One method to attempt this achievement is by bringing together neuroscientists of various orientations to present their work and to discuss their ideas together. With such an aim in mind we present this current volume based on the symposium which we organized. We hope that others will continue to face this challenge.

References

1 Adams, F.A.: The genuine works of Hippocrates (Williams & Wilkins, Baltimore 1939).
2 Aghajanian, G.K.: Influence of drugs on the firing of serotonin-containing neurons in brain. Fed. Proc. *31:* 91–96 (1972).
3 Bunney, W.E.; Goodwin, F.K.; Murphy, D.L.; House, K.M.; Gordon, E.K.: The 'switch process' in manic-depressive illness. Archs gen. Psychiat. *27:* 304–309 (1972).
4 Hill, D.: Historical review; in Reynolds, Trimble (eds.), Epilepsy and psychiatry, pp. 1–11 (Churchill Livingstone, Edinburgh 1981).
5 MacLean, P.D.: The triune brain, emotion of scientific bias; in The neurosciences, 2nd Study Program, pp. 336–349 (Rockefeller University Press, New York 1970).

6 Okuma, T.; Kishimoto, A.; Inone, K.; Matsumoto, H.; Ogura, A.; Matsushita, T.; Nakao, T.; Ogura, C.: Anti-manic and prophylactic effects of carbamazepine (Tegretol) on manic-depressive psychosis. Folia psychiat. neurol. jap. *27:* 283–297 (1973).
7 Papez, J.W.: A proposed mechanism of emotion. Archs neurol. Psychiat. *38:* 725–743 (1937).
8 Scheibel, M.E.; Crandall, P.H.; Scheibel, A.G.: The hippocampal-dentate complex in temporal lobe epilepsy. Epilepsia *15:* 55–80 (1974).
9 Stephan, H.: The allocortex (Springer, Berlin 1975).
10 Stevens, J.R.: Psychomotor epilepsy and schizophrenia: a common anatomy? in Brazier, Epilepsy, its phenomena in man, pp. 189–214 (Academic Press, New York 1973).
11 Trimble, M.R.: The limbic system; in Reynolds, Trimble (eds.), Epilepsy and psychiatry, pp. 217–227 (Churchill Livingstone, Edinburgh 1981).
12 Yakovlev, P.: Motility, behaviour and the brain. *107:* 313–335 (1948).

W.P. Koella, MD, Friedrich Miescher Institute, CH-4002 Basel (Switzerland) and
M.R. Trimble, MRCP, FRCPsych., Institute of Neurology, London (England)

Subject Index

Acetylcholine 29
Affective disturbances 7
– disorder 140
Aggression 18, 28, 59, 61, 62, 113
Aggressive behavior 16, 60, 61, 114, 146
Alcohol withdrawal 147
Amygdala 109, 110, 113
– kindling 120
Antimanic efficacy of carbamazepine 124
Anxiety 102
Attack behavior 17
Aura 107
Autonomic activity 27

Behaviour disturbances IX
Beta-agonists 18
Brain recordings 106

Carbamazepine 72, 73, 117, 121, 142, 160
– responders 139
Carbamazepine-10, 11-epoxide 136
Catatonic 2
Catecholamine 71, 119, 146, 160
Catego 99, 101, 103
Cholinergic 58, 146
– influences 47
Classification in psychiatry 1
Cognitive map theory 23
Cortical spiking 107

Delusions 100
– of persecution 6
Diabetes insipidus 132, 141
Diphenylhydantoin 72, 127, 142
Dominant hemisphere 89
– temporal 93
Dopamine 47, 159, 161
Dopaminergic pathways 18

Electroconvulsive therapy 144
Enkephalin 29
Entorhinal region 41
Epilepsy 58
Epileptic personality VII, 56, 75
– psychosis 5–7
Epileptiform discharges 64

Fear 22, 28
Fibrillary gliosis 159
First-rank symptoms 3, 94, 100, 104
Focal fibrillary gliosis 68
Forced normalisation 6, 70, 145

GABA 20, 45, 46, 137, 143

Habituation 49
Hallucinations 3, 6, 72, 102
Hallucinatory psychosis 9
– states 7
Hamartoma 68, 95, 159
Hebephrenic 2
Hippocampus 109, 110, 113

Subject Index

Histamine 19
Homovanillic acid 133
Humoral transmitting mechanisms 13

International Classification 4

Kindling IX, 145–147
– models 121
– seizure models 142
Klüver-Bucy syndrome 15, 24

Lateralisation 8
Laterality 67, 83, 85, 86, 90, 103, 104
Learning 141
Left temporal lobes 119
Lidocaine 146
– seizures 146
Limbic seizures 29
Lithium carbonate 125
LSD-25 160

Measurement of psychiatric symptoms 98
Memory 22, 141

Naloxone 71
Neuropsychophysiological procedures 13
Neurotransmitter 27, 143, 160
– systems 137
Noradrenaline 23, 47

Opiate 137
– receptors 29
Opioid receptors 48
Opioids 137
Oral sign 20
– tendencies 20
Overlap of symptoms VIII

Paranoia 2, 72
Paranoid 2
– syndromes 8
Perforant path 41
Personality disturbance 62
Placebo substitution 131

Plasticity 48
Postictal slow activity 108
Presynaptic efficiency 43
Prophylactic response to carbamazepine 126
Psychomotor 58
– activity 119
– epilepsy 66
Psychotropic effects of carbamazepine 118

Reaction types 3
Recognition memory 24
Reward 21
Rhinencephalon 12
Risk factors 73–76, 159

Sacred disease 157
Schaffer collaterals 41
Schizophrenia-like 158
– psychosis 82–84, 91, 104
Schizophreniform presentation 9
Self-stimulation 21
Septal area 19
– region 108, 113
– spiking 111
Serotonin 47
Sexual behavior 19
– dimorphism 19
Simplex 2
Sinistrality 67
Sleep 26, 132
Social cohesiveness 18
– dominance 18
Socioeconomic environment 159
– groups 61
Somatostatin 137
Spike and slow-wave activity 107
– – – pattern 110
Stereotypies 25, 26, 28
Substance P 29
Symptomatological overlap 158
Synaptic summation 44
– power 43

Temporal lobe lesions VIII
– – pathology 5

Subject Index

– lobectomy 83, 95
Theta waves 26
Transmitter 14, 50, 115

US-UK Diagnostic Project 4

Valproate 72
Vasopressin 25, 138, 141

Vigilance 26
Violence 109, 113

Wechsler Adult Intelligence Scale 60
White blood cell count 134
Wing Present State Examination 99
Withdrawal syndrome 132